5

Progress in

Clinical Biochemistry and Medicine

Regulation of Plasma Low Density Lipoprotein Levels

Biopharmacological Regulation of Protein Phosphorylation

Calcium-Activated Neutral Protease

Microbial Iron Transport

Pharmacokinetic Drug Interactions

With Contributions by
W. R. Bartle, V. Braun, J. M. Dietschy,
Y. Emori, M. Hagiwara, H. Hidaka,
S. Imajoh, H. Kawasaki, J. B. Meddings,
S. Ohno, K. Suzuki, S. E. Walker,
G. Winkelmann, N. E. Winslade

With 59 Figures

Springer-Verlag Berlin Heidelberg NewYork
London Paris Tokyo

ISBN-13: 978-3-642-72904-1 e-ISBN-13: 978-3-642-72902-7
DOI: 10.1007/978-3-642-72902-7

Softcover reprint of the hardcover 1st editio 1988

Typesetting and Printing: Th. Müntzer, GDR: Bookbinding: Lüderitz & Bauer, Berlin
2152/3020-543210

Editorial Board

Table of Contents

Regulation of Plasma Low Density Lipoprotein Levels: New Strategies for Drug Design

Jonathan B. Meddings and John M. Dietschy

University of Texas Health Science Center at Dallas, Southwestern Medical School, Department of Internal Medicine, 5323 Harry Hines Boulevard Dallas, Texas/USA

Hypercholesterolemia and, more specifically, high circulating levels of LDL-cholesterol have been linked to several important diseases in man. For this reason, a concerted effort is now underway to develop effective strategies to decrease the concentration of this lipoprotein in the plasma. Over the last few years, new advances have been made in our understanding of the factors that control the plasma LDL-cholesterol concentration. It is now possible to predict the changes that will occur in the plasma LDL-cholesterol concentration given any change in LDL receptor number or affinity or in the LDL-cholesterol production rate. From an analysis of these kinetic relationships several important conclusions emerge. First, in both man and the experimental animal the rate of LDL-cholesterol production is the most significant determinant of the plasma LDL-cholesterol concentration. Second, receptor-dependent LDL-cholesterol degradation accounts for the majority of total LDL-cholesterol degradation in all species studied. Third, in quantitative terms, receptor-dependent LDL-cholesterol degradation is restricted to the liver and, to a lesser extent, the small intestine. It is also clear that in order to significantly change the plasma LDL-cholesterol concentration by altering the rate of receptor-dependent LDL-cholesterol clearance, a change in receptor number, rather than receptor affinity, must be made since the system is resistant to even large variations in receptor affinity.

Progress in Clinical Biochemistry
and Medicine, Vol. 5

1 Introduction

Cholesterol is essential for normal cellular function. It is found in all cells of the body and is a major constituent of cell membranes. In addition, certain specialized tissues use cholesterol as the basic substrate for the synthesis of more complicated sterols such as bile acids and the steroid hormones. Like most lipids cholesterol has a very low solubility in water. Therefore, in order for these molecules to be transported between different organs of the body a series of lipid carriers known as lipoproteins has evolved. Over the last several years epidemiological evidence has accumulated suggesting a strong relationship between the risk of atherosclerosis and high plasma concentrations of one of these lipoproteins, low density lipoprotein (LDL) [1]. Thus, in populations with high plasma concentrations of LDL-cholesterol a correspondingly high incidence of vascular disease is seen. Since it has been demonstrated that reducing the steady-state plasma LDL-cholesterol concentration produces a proportional decrease in the risk of developing atherosclerosis [2], a major effort is now underway to develop safe and effective lipid lowering agents. In the past, these efforts have been hampered by a lack of knowledge concerning the physiological regulation of the plasma LDL-cholesterol concentration. However, recently there have been several major advances in these areas and it is now possible to precisely determine those factors that have the greatest effect on either increasing or decreasing the plasma LDL-cholesterol concentration. Thus, for the first time, it is now possible to design new therapeutic agents that are targeted at a specific aspect of LDL-cholesterol metabolism and to predict the ultimate effect these agents will have on the plasma LDL-cholesterol concentration. This chapter will briefly review the current state of knowledge concerning LDL metabolism and then focus on the factors that regulate the plasma LDL-cholesterol concentration.

2 Lipoprotein Physiology — An Overview

The general features of lipoprotein metabolism are now fairly well understood and are illustrated schematically in Fig. 1. Both dietary cholesterol and fatty acids are absorbed passively in the proximal small intestine at rates dictated by the permeability properties of the jejunal microvillus membrane [3,4]. Once inside the enterocyte, acyl-CoA: cholesterol acyltransferase rapidly esterifies free cholesterol to long-chain length fatty acids with the resultant formation of very hydrophobic cholesterol esters [4]. These are incorporated into the chylomicron (CM) particle which is then extruded into the intestinal lymph. During its synthesis in the intestine, the CM acquires several apoproteins which are important in guiding its future metabolism. Some of these, e.g., apoprotein B_{48}, are synthesized in the intestine while others, e.g., apo C and apo E, are synthesized elsewhere and are probably transferred to the CM particle in the intestinal lymph and plasma from high density lipoproteins (HDL). From the lymphatic system the CM enters the plasma and ultimately is recognized by the enzyme lipoprotein lipase which is located in the capillaries of muscle and fat. Both the ad-

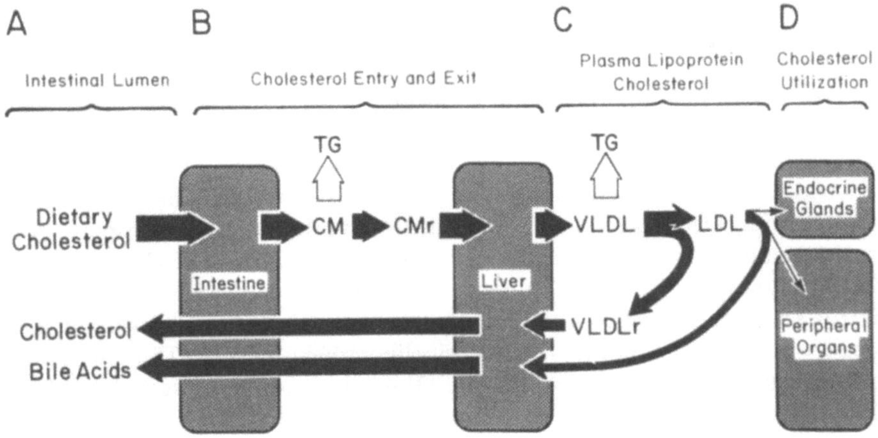

Fig. 1. General scheme for the movement of cholesterol within the body. For convenience the body has been divided into four separate compartments, labeled A through D, which represent the major components involved in the metabolism of cholesterol. The shaded regions represent the organs that participate in this process and include the liver, intestine and peripheral organs. Dietary cholesterol enters the body and is transported within the chylomicron particles (CM) which, in the periphery, are transformed into chylomicron remnants (CMr). This conversion is accompanied by the loss of triglyceride (TG) from the core of the particle. In a similar fashion the liver exports cholesterol and triglyceride to the periphery in the form of very low density lipoprotein particles (VLDL). With the loss of triglyceride these are converted either to VLDL remnants (VLDLr) or to LDL particles which are then removed from the circulation, primarily by the liver. Finally, excretion of cholesterol from the body is accomplished in the liver either by conversion of cholesterol into bile acids or by direct secretion into the bile

herence of the CM to this enzyme and subsequent enzyme activation are dependent upon apo C-II. Once activated, lipoprotein lipase catalyzes the lipolysis and removal of dietary triglyceride contained within the core of the CM particle which results in the formation of a smaller, but denser, particle termed the chylomicron remnant (CMr). During this process apo C-II is lost, lipolysis ceases and the CMr is released into the circulation from which it is cleared into the liver by a specific, high affinity remnant receptor [5]. Therefore, the net result of this pathway is the delivery of dietary triglyceride largely to the peripheral organs and dietary cholesterol primarily to the liver.

In the liver cholesterol derived from the CMr can regulate the rate of hepatic sterol synthesis. Thus, under conditions where dietary cholesterol intake is high the liver can compensate by decreasing its rate of cholesterol synthesis, thereby maintaining total sterol balance constant. The liver also secretes a triglyceride-rich particle, very low density lipoprotein (VLDL), which serves to transport triglyceride from the liver to the periphery. VLDL particles also contain apoproteins, most notably apo C, apo B_{100}, and apo E. In the periphery apo C-II regulates the recognition and partial degradation of VLDL particles in a fashion similar to that for the CM particle. However, unlike the CM, there appears to be two potential fates for the VLDL particle. First, a VLDL remnant particle (VLDLr) may be formed that is rapidly cleared by the liver. This sequence of events accounts for the majority of VLDL metabolism in species such as the rat. However in other species, including man, a significant fraction of VLDL is transformed by a poorly understood mechanism into LDL. Recent evidence suggests that at least one factor that is important in determining which transformation takes place is the relative proportion of apo E and apo B in the VLDL particle [6]. In any event, it appears that LDL is produced primarily from the catabolism of VLDL, although it should be mentioned that direct secretion of LDL from the liver may also occur. Once formed, LDL is removed from the circulation by virtually all organs of the body although, once again, the liver predominates [7].

Therefore, in summary, cholesterol enters the body from either the diet or from de novo synthesis. Due to its insolubility in water it is carried in the plasma in a variety of lipoproteins that originate either in the intestine or liver. The origin of LDL-cholesterol is not entirely clear, but the best evidence to date suggests that it is formed as a by-product of VLDL catabolism.

3 LDL-Cholesterol Catabolism

3.1 General Principles of Transport

Two factors control the steady-state concentration of LDL-cholesterol in the plasma. These include, first, the rate of inflow of LDL into the plasma space (LDL production) and, second, the rate of efflux of LDL from the plasma (LDL degradation). However, this is a deceptively simple statement for, as noted above, not only is the location of LDL production unclear but the mechanisms that control this process are poorly understood. On the other hand, while the transport processes that bring about the degradation of LDL-cholesterol are also complex, they are now fairly well worked out and are presented schematically in Fig. 2.

LDL-cholesterol, like many other particles, can cross cellular membranes by two entirely different routes. With the pioneering work of Brown and Goldstein [8], it has become clear that there exists a receptor on the surface of certain cell types that avidly binds the LDL particle. The LDL receptor is synthesized in the rough endoplasmic reticulum of these cells and subsequently inserted into the plasma membrane. Within the membrane, these receptors exhibit relatively unrestricted lateral mobility and tend to congregate in coated pits which eventually become endocytotic vesicles. Once endocytosis of the coated pit has occurred, LDL-cholesterol is separated from the receptor by progressive acidification of the vesicle contents. The free LDL receptor is then returned to the plasma membrane where it can be reused while the LDL particle undergoes further degradation [8].

A few points should be made about the physiology of this pathway. First, since the number of LDL receptors is finite there is a maximal achievable rate of transport that occurs when all receptors have become saturated. Therefore, this form of LDL degradation demonstrates saturation kinetics. Second, since the LDL receptor is expressed in only a limited number of cell types this pathway for LDL degradation is found in only certain tissues [7].

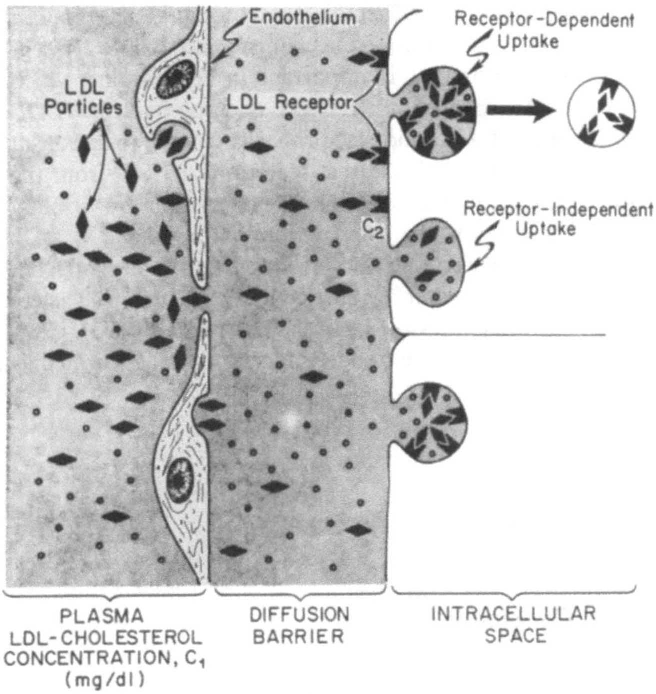

Fig. 2. Mechanisms of LDL transport into individual cells. LDL particles in the plasma are separated from the cell surface by both the vascular endothelium and the extracellular space which, together, constitute a diffusion barrier. Therefore, as discussed more completely in the text, the concentration of LDL-cholesterol in the plasma (C_1) is greater than the concentration at the surface of the cell (C_2). The LDL receptor, located on the surface of the cell, mediates receptor-dependent LDL uptake by an endocytotic process. Receptor-independent LDL uptake also presumably occurs by an endocytotic process, but this transport mechanism is poorly understood

The second pathway by which LDL-cholesterol can enter the cell is termed the receptor-independent route and this is physiologically distinct from the receptor-dependent pathway in two major respects. First, it appears to be present in all cell types and, second, the rate of LDL uptake by this route is a linear function of the plasma LDL concentration [9,10]. Thus, at low plasma LDL-cholesterol concentrations receptor-dependent uptake predominates, but with increasing plasma concentrations of LDL-cholesterol the receptor-dependent pathway becomes progressively saturated and, therefore, the receptor-independent pathway assumes greater importance. The exact molecular mechanisms of receptor-independent uptake are obscure but it has been proposed to occur by non-specific endocytosis as depicted in Fig. 2.

3.2 Degradation of LDL by Individual Organs

It is now possible to accurately measure the rate of LDL-cholesterol degradation by either of these pathways in the experimental animal. Briefly, these experiments are performed by labeling LDL in vitro and then injecting it into the animal [7,9]. Homologous LDL will be removed from the circulation by both pathways. Therefore, in order to assess each pathway independently the experiment is performed twice, once with homologous LDL and a second time with LDL that has been modified to such an extent that it no longer interacts with the LDL receptor. For this purpose it is common to use heterologous LDL that has been reductively methylated, since this reaction has been shown to be both irreversible in vivo and to result in a particle that has little, if any, affinity for the LDL receptor [9,10]. Thus, the rate of tissue accumulation of reductively methylated heterologous LDL-cholesterol reflects the activity of the receptor-independent pathway and the difference between the rate of uptake of this particle and the rate of accumulation of homologous LDL-cholesterol reflects the activity of the receptor-dependent pathway. In both cases the results can be expressed as either a true uptake rate with the units of mass of LDL-cholesterol taken up per hour per gram of tissue or as a clearance term where the same information is given as the ml of plasma completely cleared of LDL-cholesterol per hour per gram of tissue. Such data can now be easily obtained for all organs in the experimental animal.

In Fig. 3 the results of such experiments in three animal species are presented. It is important to note that the rates of uptake in this case are expressed as the μg of LDL-cholesterol taken up and degraded per h per g wet weight of tissue. There are several major conclusions to be made from such data. First, in these species LDL-

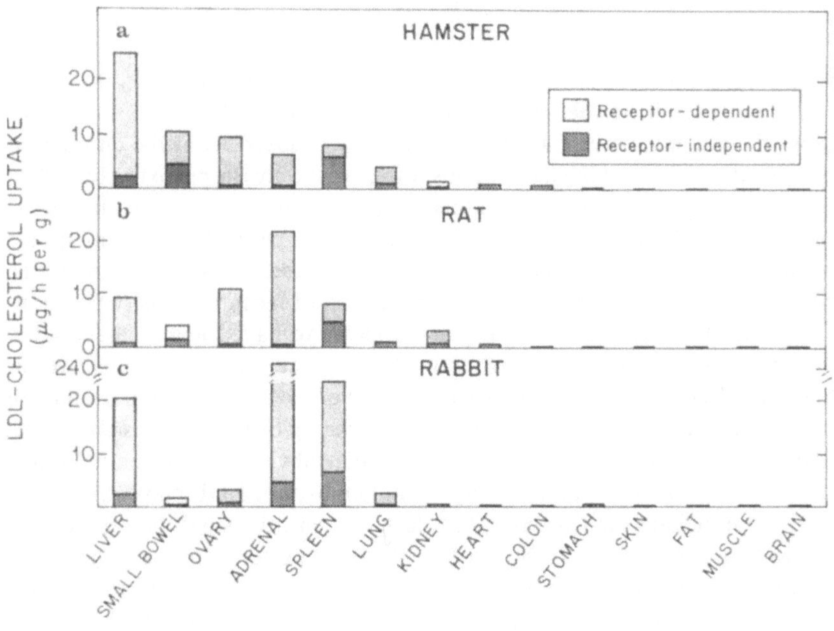

Fig. 3. Rates of LDL degradation per g of tissue in the hamster, rat and rabbit. Panels a, b and c illustrate the rates of LDL-cholesterol uptake in 14 major organs of the hamster, rat, and rabbit. Total uptake is expressed as the μg of LDL-cholesterol removed from the circulation per h per g of organ and is divided into the receptor-dependent and receptor-independent components

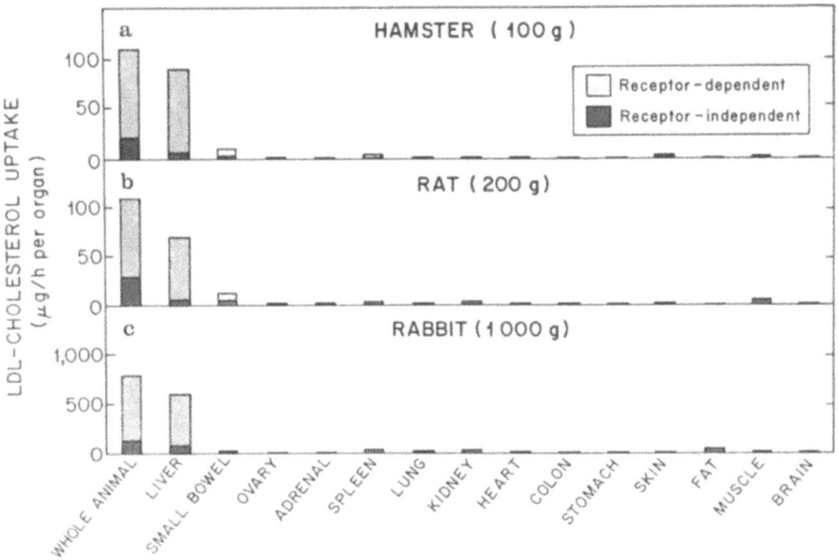

Fig. 4. Rates of LDL degradation in the whole animal and whole organ in three species. Panels a, b and c represent the same data shown in Fig. 3 but expressed as the μg of LDL-cholesterol removed from the circulation per h per whole organ. The rate of total LDL-cholesterol degradation in each animal is also illustrated and labeled as whole animal. Again the receptor-dependent and receptor-independent components of LDL uptake are shown

cholesterol is taken up in significant amounts by only a few organs of the body, primarily the liver, small bowel, spleen and endocrine organs. Second, in most of these organs the vast majority of LDL-cholesterol is removed from the circulation by the receptor-dependent pathway. For instance, in the liver approximately 90% of total uptake is receptor-dependent and this has been found to be true in all species. The only exception to this is the spleen where a much higher percentage of total uptake is receptor-independent. Since the data in Fig. 3 are expressed per gram of organ, the rate of receptor-dependent uptake is in fact a reflection of receptor density per gram of tissue. Thus, in the rabbit the adrenal gland appears to have the greatest number of LDL receptors per gram of tissue while in the hamster it is the liver. However, the weight of the adrenal glands is only a small fraction of the total weight of the animal and, therefore, when the same data are expressed as the μg of LDL-cholesterol taken up per hour per organ, as shown in Fig. 4, it becomes obvious that essentially only two organs contribute in a meaningful way to the removal of LDL-cholesterol from the plasma. These include the liver and, to a lesser extent, the small bowel. As also shown in Fig. 4, approximately 70–80% of the LDL-cholesterol that is degraded in the whole animal each hour is degraded by the receptor-dependent pathway in each species and in all cases the liver accounts for the vast majority of this uptake. Therefore, although the adrenal gland has a high receptor density, its total contribution to LDL-cholesterol removal is minimal, because of its small mass.

An additional point that must be made is that all measurements presented in Figs. 3 and 4 have been made at a single, fixed plasma LDL-cholesterol concentration. Therefore, only under these experimental conditions can it be concluded that the liver of these three species removes approximately 60–75% of the LDL-cholesterol produced each hour. If the rate of LDL-cholesterol production were to increase, and therefore elevate the plasma LDL-cholesterol concentration, the relative contribution of each organ to total LDL uptake would change. In order to accomodate such variations in a model of LDL-cholesterol metabolism it is necessary to first construct a theoretical framework describing the relationships that exist between the plasma LDL-cholesterol concentration and its determinants. Second, a series of experiments must then be performed to measure the values of the parameters used to describe the theoretical model. Once completed, such a model would then allow the prediction of the plasma LDL-cholesterol concentration given any change in the LDL-cholesterol production rate or the rate of LDL-cholesterol degradation by either pathway in any organ. In fact such a model now exists, and the experimental work to derive the kinetic parameters of LDL-cholesterol metabolism has been completed in four species including the rat, hamster, rabbit and man. The derivation of this model is described in the following section.

4 Kinetics of LDL-Cholesterol Regulation

4.1 Development of a Theoretical Model

As noted above, the steady-state LDL-cholesterol concentration in the plasma is directly determined by the relative rates of LDL-cholesterol production and LDL-cholesterol degradation. Although the location of LDL-cholesterol production remains unclear, the rate of LDL-cholesterol production can be defined by a single rate constant. As illustrated in Fig. 1 this constant is probably determined by at least two prior steps, the rate of hepatic VLDL secretion and the rate of conversion of VLDL to LDL.

The rate of LDL-cholesterol removal is a more complicated process since this occurs by two independent methods in multiple organs. However, this problem can be addressed mathematically in the following manner. In any organ the rate of receptor-independent uptake (J_i) is a linear function of the LDL-cholesterol concentration at the plasma membrane (C_2) and, therefore, is described by the following relationship [11, 12]

$$J_i = PC_2 \tag{1}$$

where P is the proportionality constant of this relationship and is analogous to a membrane permeability coefficient. In a similar manner the rate of receptor-dependent uptake (J_d) can be described by Michaelis-Menton kinetics [11, 12]

$$J_d = \frac{J^m C_2}{K_m + C_2} \tag{2}$$

where J^m is the maximal achievable transport velocity (a reflection of receptor number) and K_m is the concentration at which one-half maximal transport rates occur (a reflection of receptor affinity).

There are two important points to be made regarding these equations. First, the K_m value is not identical to the binding constant (K_b) that is obtained with LDL binding studies since the former term (K_m) is defined as the concentration of LDL-cholesterol required to obtain one-half maximal uptake rates, not one-half maximal binding rates, as is the case for K_b. Therefore, unless 100% of bound LDL-cholesterol is internalized, the value of K_m must necessarily be greater than the value of K_b obtained in binding studies. Secondly, the concentration of LDL-cholesterol at the membrane (C_2) is not necessarily equal to the concentration of LDL-cholesterol in the plasma (C_1). In fact, C_2 will always be less than C_1 by an amount that is proportional to the resistance the LDL particle encounters in moving from the plasma space across the extracellular fluid to the plasma membrane of the cell (Fig. 2). The relationship between C_1 and C_2 can be described by the following equation [11, 12].

$$C_2 = C_1 - J_t R \tag{3}$$

In this equation J_t represents the total rate of LDL-cholesterol uptake, i.e., J_d plus J_i, and R represents the resistance of the diffusion barrier separating the plasma space from the cellular membrane. Thus, in situations where either diffusion barrier resistance (R) or the rate of LDL removal from the extracellular space (J_t) are high, C_2 will be significantly lower than C_1.

Therefore, in any organ the total rate of LDL-cholesterol uptake (J_t) can be described by the following relationship

$$J_t = J_d + J_i = \frac{J^m C_2}{K_m + C_2} + PC_2 \tag{4}$$

Since C_2 is not an easily measurable quantity it is preferable to restate this relationship in terms of the plasma LDL-cholesterol concentration (C_1) using equation 3. This results in Eq. (5) [12]

$$J_t = \frac{\left[\begin{array}{r}(K_m + C_1 + PRK_m + 2PRC_1 + J^m R) - \sqrt{(K_m + C_1 + PRK_m + 2PRC_1 + J^m R)^2} \\ \overline{-4[PR(PRC_1K_m + PRC_1^2 + RJ^mC_1 + C_1K_m + C_1^2) + J^mC_1R]}\end{array}\right]}{2R(PR + 1)} \tag{5}$$

Although Eq. (5) has a rather complicated form, it accurately describes the relationship that exists between the plasma LDL-cholesterol concentration and the rate of LDL-cholesterol uptake in any organ. Total body LDL-cholesterol uptake is, of course, the summation of individual uptake rates from all organs of the body.

4.2 *The Kinetic Parameters of LDL-Cholesterol in the Experimental Animal*

It can now be seen that the rate of LDL-cholesterol uptake in any organ at any plasma LDL-cholesterol concentration is determined by four parameters, J^m (a reflection of receptor number), K_m (a reflection of receptor affinity), P (the permeability coefficient describing receptor-independent LDL uptake) and R (the diffusion barrier resistance). By actually measuring J_t in individual organs over a wide range of plasma LDL-cholesterol concentrations, it is possible to fit a curve, of a shape described by Eq. (*5*), to the data by computerized nonlinear regression analysis. In so doing, the best-fit values for all four parameters can be obtained for each individual organ. These experiments have now been performed in essentially all organs of both the rat and the hamster at 20–30 different plasma LDL-cholesterol concentrations [12]. Representative curves obtained from experiments in the hamster are shown in Fig. 5. These curves represent the best-fit relationship between the experimentally determined plasma LDL-cholesterol concentration and the rate of total LDL-cholesterol uptake obtained in the liver, small bowel, spleen and kidney. The general shape of these curves is dictated by Eq. (*5*), but the shape of the particular curve for any organ is specified by the values for each of the four parameters appropriate for that tissue. Rates of LDL-cholesterol uptake in all other organs are too low to be expressed on the same scale, but if the rate of LDL-cholesterol uptake in all organs is summed over the range of plasma LDL-cholesterol concentrations, then the rate of LDL-cholesterol uptake in

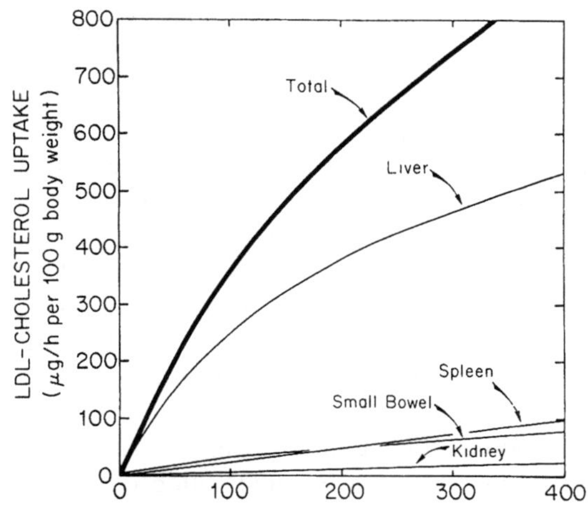

Fig. 5. Kinetic curves for LDL transport (degradation) in the whole hamster. The curves shown represent the μg of LDL-cholesterol taken up and degraded each h by both the receptor-dependent and receptor-independent routes in 4 major organs of the hamster at varying concentrations of plasma LDL-cholesterol. The summation of such curves for all organs gives the rate of degradation of LDL-cholesterol by the whole hamster (Total) as a function of the plasma LDL-cholesterol concentration

the whole animal at any plasma concentration of LDL-cholesterol is obtained and is shown in Fig. 5 as the curve labelled Total.

There are some very important conclusions to be derived from these types of experiments. First, receptor-dependent uptake can be detected in several organs of both the rat and the hamster, and include the liver, adrenal gland, ovary, spleen, and small bowel. However, in quantitative terms only the liver and small bowel are important for the receptor-dependent removal of LDL-cholesterol from the plasma. Second, although the receptor number per gram of tissue, as expressed by the maximal achievable transport rate (J^m), varies widely in these organs from 5.5 to 126 µg per g, the K_m value for the LDL receptor in vivo is remarkably constant, averaging about 90 mg/dl in all organs of both the rat and the hamster. Thus, the receptor-dependent transport system will not become saturated until the plasma concentration of LDL-cholesterol is well in excess of 400 mg/dl. Third, these experiments allow direct calculation of R, the diffusion barrier resistance, in each organ of the experimental animal. In both the rat and hamster, the values obtained for this parameter are so low, relative to the total rate of LDL-cholesterol uptake (J_t), that C_2 essentially equals C_1 in all organs where this could be determined. In this case where the value of R approaches zero, equation 5 reduces to the much simpler form

$$J_t = \frac{J^m C_1 + PC_1K_m + PC_1^2}{K_m + C_1} \tag{6}$$

where J_t is now a function of the plasma LDL-cholesterol concentration (C_1) and is modulated by only three parameters, receptor number (J^m), receptor affinity (K_m) and the LDL-cholesterol permeability coefficient (P). There is one final point that warrants emphasis. Under steady-state conditions, where the plasma LDL-cholesterol concentration remains constant, total body LDL-cholesterol uptake (J_t) must equal the production rate of LDL-cholesterol. Therefore, in Fig. 5 the vertical axis represents not only the rate of total LDL-cholesterol uptake but also the LDL-cholesterol production rate. In this case the curve labeled Total describes the relationship that exists between the rate of LDL-cholesterol production and the plasma LDL-cholesterol concentration under conditions where LDL receptor number (J^m) and affinity (K_m) and the permeability coefficient (P) remain constant. Thus, it can be seen that a LDL-cholesterol production rate of approximately 130 µg/h produces a plasma LDL-cholesterol concentration of 25 mg/dl in the normal hamster. If the rate of LDL-cholesterol production were to increase to 370 µg/h without a change in J^m, K_m or P then the plasma LDL-cholesterol would rise to 100 mg/dl in a manner described by the curve labeled Total in Fig. 5.

Thus, in the experimental animal it is now possible to measure all the parameters that determine the plasma LDL-cholesterol concentration. Receptor number, receptor affinity, LDL-cholesterol permeability and diffusion barrier resistance can each be measured directly in all organs and the LDL-cholesterol production rate can be measured in the whole animal. In intact man it is, of course, impossible to perform the same type of experiment. However, with the knowledge gained from the experimental animal it is now possible to quantitate the same parameters in man and, therefore, make accurate predictions of human plasma LDL-cholesterol concentrations given any change in these transport parameters. This can be accomplished using the model derived from animal studies and data obtained from published human LDL turnover studies [13]. The derivation of these parameters in man is discussed in the next section.

4.3 The Kinetic Parameters of LDL-Cholesterol in Intact Man

Although it is impossible to derive the human kinetic constants for LDL metabolism in individual organs, it is possible to obtain an average value for each parameter that pertains to the whole body. This can be done using the theoretical model presented above and by making certain assumptions found to be true in the animal model. There are four parameters to be considered and each will be dealt with in turn.

First, where it could be directly calculated, diffusion barrier resistance in experimental animals was negligible and, therefore, the plasma concentration of LDL-cholesterol appears to be a good approximation of the LDL-cholesterol concentration at the plasma membrane. Thus Eq. (*6*) closely approximates the relationship that exists between the plasma LDL-cholesterol concentration and the rate of LDL-cholesterol uptake in either individual organs or the body as a whole.

Of the remaining three parameters P, the permeability coefficient for receptor-independent LDL uptake, is the most easily derived. Since this represents the slope of the relationship between the plasma LDL-cholesterol concentration and the rate of receptor-independent LDL-cholesterol transport, it is apparent that an accurate measurement of the receptor-independent uptake rate at a well defined plasma LDL-cholesterol concentration would determine the value of this parameter. This can be most easily accomplished by examining the LDL turnover data obtained in patients with homozygous familial hypercholesterolemia, since virtually all LDL-cholesterol uptake in these patients occurs by the receptor-independent pathway [14]. From an analysis of published series involving LDL turnover studies in patients with this disorder [13] it can be calculated that the value of P in intact man is approximately 0.003 mg/h per kg per mg/dl. The definition of this value is extremely important since not only does it describe the rate of receptor-independent LDL-cholesterol uptake but it also fixes the slope of the curve relating total LDL-cholesterol uptake to the plasma LDL-cholesterol concentration under conditions where the receptor-dependent system has become saturated.

To obtain reasonable estimates for the two remaining parameters, receptor number (J^m) and receptor affinity (K_m), a different procedure must be used. Since the value of P defines the slope of the relationship that exists between the plasma LDL-cholesterol concentration and the rate of total LDL-cholesterol uptake (J_t) at high plasma LDL-cholesterol concentrations, nonlinear regression techniques can be used to estimate values for the remaining two parameters, J^m and K_m, if the rate of total LDL-cholesterol uptake can be measured over a fairly wide range of plasma LDL-cholesterol concentrations. However, it is mandatory that these relatively high plasma LDL-cholesterol concentrations not be secondary to a change in either J^m or K_m since estimates for the values of these parameters derived from such data would not represent the true "normal" value for these parameters.

There are a large number of LDL turnover studies in the literature that evaluate the rate of total LDL-cholesterol uptake in ostensibly normal young individuals. The rates of LDL-cholesterol uptake in this group provide the first series of points that can be fixed. Ideally, the second series of points should be as far as possible from the first. However, it is difficult to find normal young people with high plasma concentrations of LDL-cholesterol that can be confidently ascribed to overproduction alone. However, there are a group of patients with a known receptor number and a relatively high average plasma LDL-cholesterol concentration that can be conveniently used for this purpose. These are the familial hypercholesterolemic heterozygotes who appear to express one-half the normal number of LDL receptors and who have an average plasma LDL-cholesterol concentration of 230 mg/dl [14]. Using the value of P obtained above and assuming that these patients express exactly one-half the normal receptor number, the rate of total LDL-cholesterol uptake that would have occurred at that plasma LDL-cholesterol concentration if a normal complement of LDL receptors were present can be calculated. By combining the results from both normal subjects and this group, a large data set can be obtained that relates the total LDL-cholesterol uptake rate to the plasma LDL-cholesterol concentration over a range that varies from 58 mg/dl to 266 mg/dl. It is possible to then fit a curve, of a shape described by Eq. (*6*), to these data points using computerized nonlinear regression techniques. Since P has already been fixed, only two further parameters need be derived, J^m and K_m. The results of this calculation reveal that the relationship between the rate of human LDL-cholesterol uptake and the plasma LDL-cholesterol concentration can be described in terms of Eq. (*6*) with a value for J^m equaling 0.78 mg/h per kg, for K_m of 90 mg/dl and for P of 0.003 mg/h per kg per mg/dl.

This results in a very powerful mathematical tool that can be used to analyze the regulation of both human and animal plasma LDL-cholesterol concentrations. It is now possible to predict the changes that will occur if, for instance, the production rate (J_t) of LDL-cholesterol were to double or a drug were to induce a two-fold increase in hepatic LDL receptors. However, before discussing these aspects of the model it is important to compare the regulation of human plasma LDL-cholesterol with that of animals, using the hamster as a model.

4.4 Comparisons Between the Kinetics of LDL-Cholesterol in Man and the Hamster

The relationship that exists between the LDL-cholesterol production rate, the rate of LDL-cholesterol degradation (either by the receptor-dependent or -independent pathways) and the plasma LDL-cholesterol concentration can be expressed in a variety of formats. Three of these are depicted in Fig. 6. In panels a and d the rate of LDL-cholesterol uptake is shown as a function of the plasma LDL-cholesterol concentration. In both the hamster and man receptor independent uptake (J_i) is a linear function of the plasma LDL-cholesterol concentration whereas the rate of total uptake (J_t) assumes the expected curvilinear relationship. It must be emphasized again that these are the "normal" curves for these two species where the value of P, J^m and K_m have their "normal" values. Therefore, the curve describing the rate of total LDL-cholesterol uptake (J_t) as a function of the plasma LDL-cholesterol concentration is also the curve defining the relationship between the steady-state LDL-cholesterol production rate and the resultant plasma LDL-cholesterol concentration if receptor

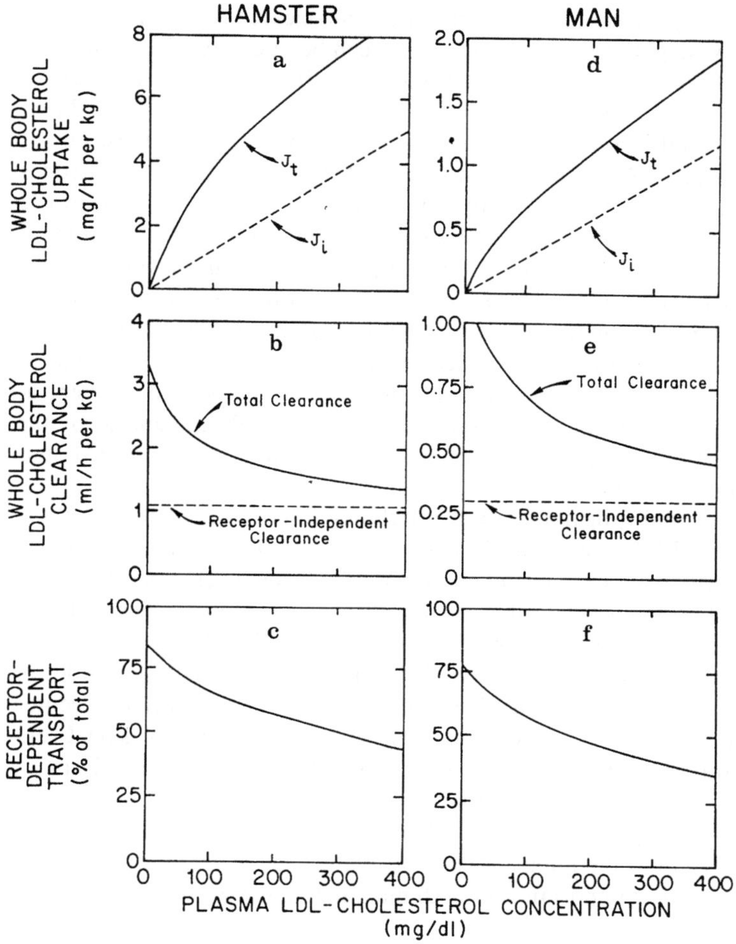

Fig. 6. Comparisons between the kinetics of LDL transport in either man or the hamster. Panels a and d illustrate the rate of total LDL-cholesterol uptake either by the receptor-independent pathway (J_i) or the sum of the receptor-independent and receptor-dependent uptake rates (J_t) in either man or the hamster. In panels b and e the same data are presented as either total or receptor-independent clearance rates representing the ml of plasma totally cleared of its LDL-cholesterol content per h per kg body weight. In panels c and f the receptor-dependent LDL-cholesterol uptake rates are presented as a percentage of total LDL-cholesterol uptake and are plotted as a function of increasing plasma LDL-cholesterol concentration

number, receptor affinity and LDL-cholesterol permeability remain constant. Thus, for any given LDL-cholesterol production rate (J_t), moving horizontally to the right to the curve labeled J_i reveals the expected increase in the plasma LDL-cholesterol concentration that would occur if all LDL receptors in the body were lost.

Most LDL turnover studies, however, are not reported in this format. Rather, the measured values obtained from such studies are the plasma LDL concentration and the fractional catabolic rate of the injected LDL, described as pools/day. This latter value is essentially a clearance term (ml of plasma cleared of its LDL content per hour per kg body weight). Such clearance values can be calculated by dividing the LDL uptake rate by the plasma LDL-cholesterol concentration at which it was measured. Thus, the same data shown in panels a and d are represented in panels b and e as the clearance rate of LDL-cholesterol. Two important points regarding these curves should be made. First, the rate of receptor-independent clearance is constant and does not vary with the plasma LDL-cholesterol concentration. Second, total clearance decreases with increasing plasma LDL-cholesterol concentrations and, importantly, this decrease occurs despite a constant LDL receptor number and affinity. Thus, the combination of a low clearance rate (low fractional catabolic rate) and an increased plasma LDL-cholesterol concentration cannot be used alone to infer a decrease in receptor number since the clearance rate (fractional catabolic rate) would be expected to decrease simply on the basis of the increased plasma LDL-cholesterol concentration. Only if the decrease seen in the clearance is greater than that predicted by the increase in the plasma LDL-cholesterol concentration can it be concluded that receptor number must have declined.

The final panels, c and f, demonstrate the declining importance of the receptor-dependent pathway for LDL removal as the concentration of plasma LDL-cholesterol increases. The similarity between man and hamster is most evident here since the curves of both species are almost superimposible. At normal concentrations of plasma LDL-cholesterol the experimental animal degrades approximately 70–80% of LDL-cholesterol by the receptor-dependent process while in man this figure is of the order of 60%. The difference between these values has been used as evidence that LDL receptor number or activity is down regulated in man. However, the normal plasma LDL-cholesterol concentration in the hamster is about 25 mg/dl while in man equals about 70 mg/dl. Therefore, if this comparison is made at a constant plasma LDL-cholesterol concentration of 25 mg/dl, both species would clear about 75% of their LDL by the receptor-dependent route. This implies that in a relative sense, receptor-dependent LDL transport is of similar importance in both man and hamster.

5 Alterations in the Plasma LDL-Cholesterol Concentration

5.1 Manipulating the Kinetic Parameters of LDL-Cholesterol

While Eqs. (*5*) and (*6*) describe quite accurately the relationship between the rate of LDL uptake and the plasma LDL concentration, it would be preferable to relate the plasma LDL concentration directly to the rate of LDL production and removal. This can be accomplished by rearranging Eq. (*6*) so that the plasma LDL-cholesterol concentration is now the dependent variable that varies with the rate of LDL production and removal in the following manner

$$C_1 = \frac{(J_t - J^m - PK_m) + \sqrt{(J_t - J^m - PK_m)^2 + 4PK_mJ_t}}{2P} \qquad (7)$$

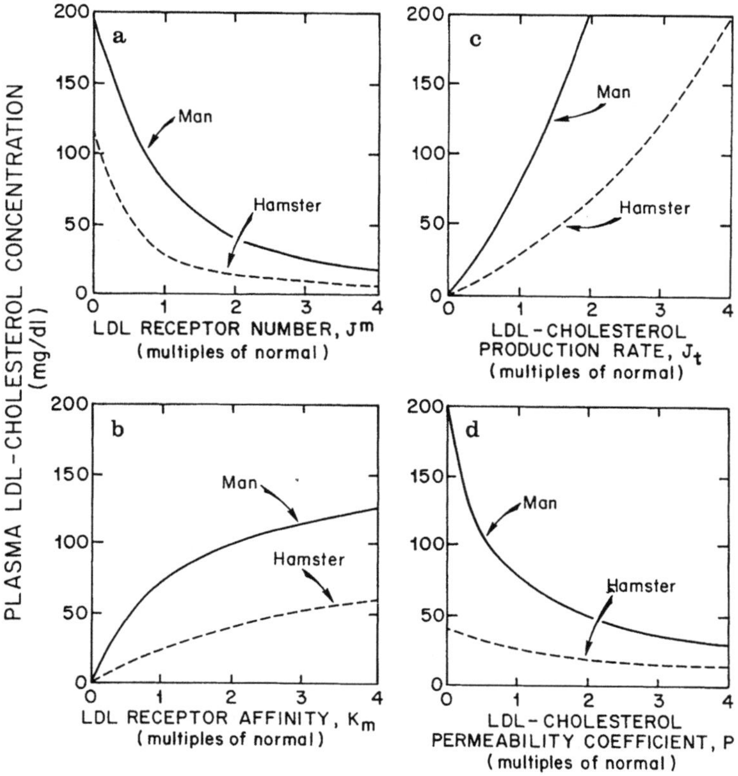

Fig. 7. Manipulation of the kinetic parameters regulating the plasma LDL-cholesterol concentration. Panels a through d illustrate the changes in the plasma LDL-cholesterol concentration expected to occur with a change in one of the four kinetic parameters. In each case the data are shown for both man and the hamster and the units for each kinetic parameter are expressed as multiples of the normal value. The normal value for LDL receptor number (J^m) in man is 0.78 mg/h per kg and in the hamster equals 4.85 mg/h per kg. Similarly, normal values for LDL receptor affinity (K_m) are 90 mg/dl in man and 91 mg/dl in the hamster. Normal LDL-cholesterol production rates (J_t) in man equal 0.54 mg/h per kg and in the hamster are 1.28 mg/h per kg. Finally the normal value for the LDL-cholesterol permeability coefficient (P) in man is 0.003 mg/h per kg per mg/dl while in the hamster this value equals 0.0123 mg/h per kg per mg/dl

Using Eq. (7) one can then examine graphically the effects of changing any parameter on the plasma LDL-cholesterol concentration. It should be mentioned that in the experimental animal where the values of K_m, J^m and P are now known in each individual organ, the overall effects of changing any one of these in a single organ can also be predicted. However, in man the analysis can only be made for the body as a whole.

Figure 7 illustrates the changes that occur in the plasma LDL-cholesterol concentration of either man or hamster when any one of the four variables is systematically altered and the remainder are kept constant. In each case the units of the variable are expressed as multiples of normal. In man the normal values are a receptor number (J^m) of 0.78 mg/h per kg, a receptor affinity (K_m) of 90 mg/dl, a permeability coefficient (P) of 0.003 mg/h per kg per mg/dl, and a production rate (J_t at the normal plasma

LDL-cholesterol concentration of 70 mg/dl) of 0.55 mg/h per kg. In the hamster the normal values are a J^m of 4.85 mg/h per kg, a K_m of 91 mg/dl, a P of .0123 mg/h per kg per mg/dl and a J_t of 1.28 mg/h per kg.

Panels a and b depict the changes that will occur in the plasma LDL-cholesterol concentration if an alteration is made in the receptor-dependent removal pathway either by drug therapy or by diet. As shown in panel a, if all receptor activity were lost, in man the plasma LDL-cholesterol concentration would increase to 183 mg/dl and in the hamster would rise to 120 mg/dl. Conversely, if some manipulation were to double LDL receptor number, the plasma LDL-cholesterol concentration in normal man would decline from approximately 70 mg/dl to 36 mg/dl.

Panel b demonstrates the changes that would occur in the plasma LDL-cholesterol concentration with an alteration in the affinity of a normal number of LDL receptors. This might occur through a change in the receptor structure, a change in the lipoprotein (particularly apo B) or the presence of competing lipoprotein particles. However, even large increases in the value of K_m have a quantitatively small effect on the plasma LDL-cholesterol concentration. For instance, in man a four-fold decrease in the affinity of the LDL receptor (to a K_m of 360 mg/dl) would only increase the plasma LDL-cholesterol concentration from 70 mg/dl to 119 mg/dl. Similarly, a two-fold increase in the affinity of the receptor has a relatively small effect and would result in a lowering of the plasma LDL-cholesterol concentration to only 48 mg/dl. Similar relationships exist in the hamster.

Panel c illustrates the effects of varying the LDL production rate, the parameter that quantitatively has the most profound effect on the plasma LDL-cholesterol concentration. For both man and hamster there exists an almost linear relationship between these two variables such that in man, doubling the LDL production rate results in an increase in the plasma LDL-cholesterol concentration from 70 mg/dl to 190 mg/dl. The curve in the hamster is similar but the relationship is less steep. Conversely, any manipulation that leads to a decline in the rate of LDL production, either by decreasing the rate of hepatic VLDL secretion or inhibiting the conversion of VLDL to LDL, will markedly decrease the plasma LDL-cholesterol concentration in an almost linear fashion.

The final parameter that theoretically can be altered is the receptor-independent pathway of LDL catabolism. Although there have been studies in man suggesting that this process may change in certain disease states, no clear-cut evidence exists for this. In the animal models, where methylated heterologous LDL preparations have been used to evaluate this pathway, no manipulation has been found to date that can alter the rate of receptor-independent LDL clearance. However, the changes that would be expected to occur with a change in the value of the LDL permeability coefficient (P) are diagrammed in panel d of Fig. 7. In the experimental animal there is little change in the plasma LDL-cholesterol concentration with large changes in the activity of the receptor-independent pathway. However, in man it would be expected that a relatively large increase in the plasma LDL concentration could occur with a four-fold decrease in the activity of this pathway. However, it is also clear from the curves shown in panel d that attempts to increase LDL clearance by this route, even if successful, would have little impact on the steady-state plasma LDL concentration.

5.2 Regulation of LDL Levels in Man and Interpretation of LDL Turnover Studies

From the data reviewed above and in Fig. 7, it is apparent that there is now available a sound theoretical basis for understanding why the plasma LDL-cholesterol level increases in certain pathological states and how pharmacological interventions might lower the levels of these circulating lipoproteins. From the practical, experimental point of view, it is now entirely possible to make the necessary measurements of LDL metabolism in the experimental animal necessary to confirm these various theoretical relationships and so gain detailed insights into how a given physiological or pharmacological maneuver alters LDL metabolism and circulating LDL- cholesterol concentrations. The ability to make such measurements in man, however, is much more limited. Nevertheless, with proper interpretation of LDL turnover studies, it is possible to obtain considerable information concerning the specific parameters of LDL metabolism that influence the plasma cholesterol levels. This section will deal more specifically with the application of the theoretical model for LDL metabolism to man and will discuss how LDL turnover data can be used to obtain specific information on such parameters as the production rate (J_t) and receptor number (J^m).

From the preceding discussion it is clear that the two most important parameters controlling the plasma LDL-cholesterol concentration in man are the number of LDL receptors present in the entire body (J^m) and the rate of LDL production (J_t). Changing the rate of receptor-independent clearance or the affinity of the receptor-

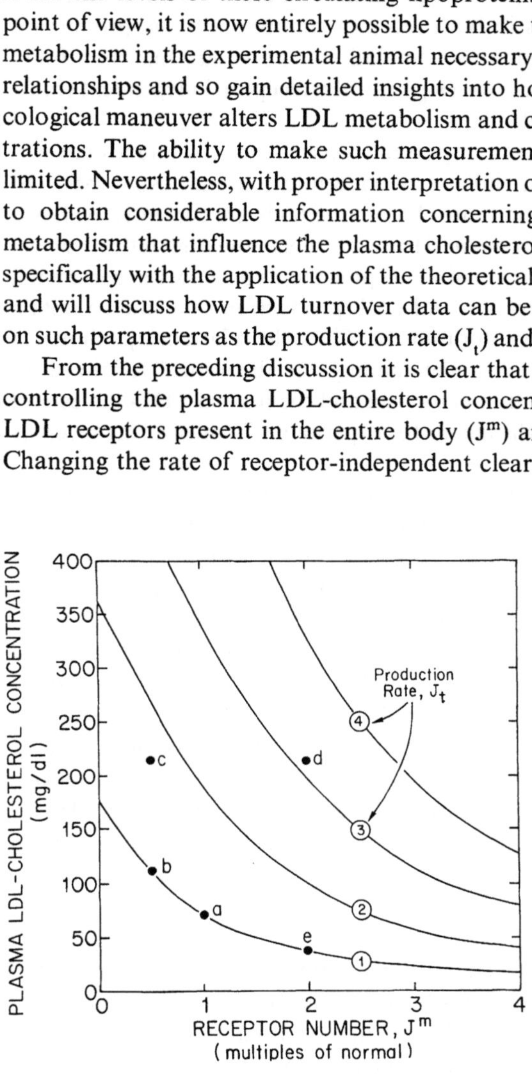

Fig. 8. Variability in human plasma LDL concentrations. This illustrates the effect of changing either the LDL receptor number (J^m) or the LDL-cholesterol production rate (J_t) on the plasma LDL-cholesterol concentration in man. The units of either kinetic parameter are once again expressed as multiples of the normal values given in the text. The significance of the points labeled a through e are discussed in the text

dependent system is not only impractical, but even large changes in these two parameters would result in only modest declines in the plasma concentration of LDL-cholesterol. In order to simultaneously evaluate the effects of both receptor number and LDL production rates, it is convenient to present these curves in the format shown in Fig. 8. In this diagram the plasma LDL-cholesterol concentration is presented as a function of the LDL receptor number (the units being multiples of normal) at four different LDL production rates (again, the units are expressed as multiples of normal). On such a graphic presentation it is possible to relate changes in the plasma LDL-cholesterol concentration to changes in both receptor number (J^m) *and* production rate (J_t). It should be emphasized that in this formulation it is assumed that the affinity of the LDL molecule for the LDL receptor (K_m) and the receptor-independent transport rate (P) have remained constant.

This formulation has one other very important consequence. In intact man it is possible to measure the plasma LDL-cholesterol concentration and, using the LDL turnover technique, the LDL production rate (J_t). No methods are currently available for quantitating receptor number in man. However, once the theoretical curves illustrated in Fig. 8 are available, it is possible to plot the two known experimental values, i.e., LDL-cholesterol concentration and LDL production rate, and so obtain graphically the third value, i.e., the receptor number. Thus, the relationships shown in Fig. 8 not only provide the means for understanding the theoretical relationships between the various parameters of LDL metabolism and the plasma LDL-cholesterol concentration, but they also provide the means for experimentally measuring receptor number in intact man.

The relationships shown in Fig. 8 provide several very important insights into the regulation of LDL-cholesterol levels in man. First, it can be seen that patients with a plasma LDL-cholesterol concentration in excess of 200 mg/dl must have a defect that involves LDL overproduction since complete loss of receptors alone will increase the plasma LDL-cholesterol concentration to only approximately 180 mg/dl. Second, although doubling LDL receptor number under normal circumstances has a quantitatively small effect on the plasma LDL-cholesterol concentration, this effect is magnified under conditions where LDL overproduction exists. Although the relative change remains constant, the absolute decline in the plasma LDL-cholesterol concentration is much greater. Third, the most efficient means of decreasing the plasma LDL-cholesterol concentration is to decrease the rate of LDL production. This is true because the response of the plasma LDL-cholesterol concentration to a given percentage decrease in production rate far exceeds the response seen to a similar percentage increase in receptor number.

The points labeled a through e in Fig. 8 illustrate several important examples that might occur in man under pathological circumstances. Point a represents the normal young human with a plasma LDL-cholesterol concentration of 70 mg/dl and both a LDL receptor number and production rate that are in the normal range. If the activity of one-half of the LDL receptors were lost, either as a result of dietary intake or disease, without a change in LDL production rate, the plasma LDL-cholesterol concentration would increase to just over 110 mg/dl (point b). Point c represents the case of the patient who is the heterozygote for familial hypercholesterolemia. In this case it is evident that in addition to having only one-half the normal complement of LDL receptors, these patients have a production rate that is 1.7 times that of normal man.

It is the elevated production rate that accounts for almost 60 % of the increase evident in their plasma LDL-cholesterol concentration. The same plasma LDL-cholesterol concentration as seen in the familial hypercholesterolemic heterozygote can also be attained under conditions where LDL receptor number has actually increased (point d). Although no clinical parallel exists for this point, it is theoretically possible to double LDL receptor number and yet have a plasma LDL-cholesterol concentration greater than 200 mg/dl, provided that there is a 3.2-fold increase in the LDL production rate. If the rate of LDL production in this case were to be reduced to the normal level (point e) then the plasma LDL-cholesterol level would fall to approximately 35 mg/dl.

Using this type of plot, it is possible to dissect the mechanisms that underlie an increase in the plasma LDL-cholesterol concentration in different clinical states. For example, it has been recognized for some time that the plasma LDL-cholesterol concentration increases with advancing age. This does not appear to occur in the experimental animal, and it has been suggested that in man the underlying mechanism is a decline in LDL receptor activity that presumably is mediated by dietary effects. However, when the data from LDL turnover studies performed in normal volunteers, with ages varying from the early teens to the mid-sixties, is plotted on Fig. 8 a different pattern emerges [13]. Although the plasma LDL-cholesterol concentration does indeed increase with age it does so in a pattern consistent with increasing LDL production, while LDL receptor number remains constant.

Thus, one important use for this type of analysis is to elucidate the physiological events that control the plasma LDL-cholesterol concentration. Another equally important use of these relationships is to examine the mode of action of various therapeutic and dietary manipulations in order to more rationally approach the treatment of hypercholesterolemia in man. For example, drugs such as mevinolin are competitive inhibitors of HMG CoA reductase, the enzyme that catalyzes the rate-limiting step in cholesterol synthesis. It is likely that their mode of action involves depletion of intracellular cholesterol stores and subsequent induction of LDL receptor activity. Two clinical trials evaluating LDL kinetics in patients treated with mevinolin have been analyzed by these techniques [13, 15, 16]. Although one of these studies does reveal the predicted increase in LDL receptor activity with mevinolin therapy [16], the second suggests that mevinolin lowered the plasma LDL-cholesterol concentration by an effect on the LDL-cholesterol production rate while leaving LDL receptor activity virtually unchanged [15]. Another example to demonstrate the usefulness of these methods is in the evaluation of the effect of dietary intake of saturated and polyunsaturated fats. It has been recognized for some time that supplementing the diet of a hamster with saturated fatty acids leads to a rapid increase in the plasma LDL-cholesterol concentration. Using these techniques in the experimental animal, it can be shown that these diets result in both a dose and time dependent suppression of LDL receptor activity that is fully reversible upon resumption of a normal diet [17]. Furthermore, it is also possible to separately evaluate the increase in LDL-cholesterol production rates seen in these animals when the diets are enriched in cholesterol. Thus, this form of analysis can provide important, and quantitative, information regarding the regulation of the plasma LDL-cholesterol concentration during dietary or pharmacological manipulations.

6 Utilizing These Models for the Design of Therapeutic Agents

With the recognition that reductions in the plasma LDL-cholesterol concentration lower the risk of developing atherosclerotic vascular disease, there has been renewed interest in the pharmacological therapy of hypercholesterolemia. Until recently, however, there has been no rational and systematic approach to the development of new pharmacological agents for lowering plasma LDL-cholesterol levels. Rather, new compounds often have come from the almost accidental observation that a particular agent happened to lower cholesterol levels when administered to animals or to man. With the availability of this new information on the quantitative physiology of LDL metabolism, this situation should change. In fact, it is now possible 1) to define in rather precise terms how new pharmacological agents should work to lower plasma LDL-cholesterol levels, 2) to identify which organs these agents should be targeted at and 3) to even predict in quantitative terms how effective such new agents should be.

As outlined above, recent analysis suggests that both increased production of LDL-cholesterol and suppression of receptor-dependent transport are the two most important factors responsible for the development of hypercholesterolemia in man. Furthermore, the kinetic analysis of LDL metabolism also indicates that the two most powerful ways in which to lower plasma cholesterol levels are to decrease the production rate of LDL (J_t) and to increase receptor number (J^m) (Fig. 8). It follows, therefore, that new strategies for the design of pharmacological agents should include compounds that alter either, or both, of these two processes.

Clearly the most important factor governing the plasma LDL-cholesterol concentration is the LDL-cholesterol production rate. The rate of this process, in turn, is determined by at least two other processes, i.e., the rate of hepatic VLDL-cholesterol secretion and the rate at which this VLDL particle is converted to LDL. At the present time, there is little known about the factors that control each of these processes; however, it is clear that if either the rate of hepatic VLDL secretion could be reduced or the cholesterol content of these particles decreased, then plasma LDL-cholesterol levels would fall dramatically. Furthermore, drugs that stimulate the conversion of VLDL particles to VLDL remnants, rather than to LDL, would have a similar dramatic effect. Essentially, for every 10% reduction in the rate of LDL-cholesterol production (J_t), a fall in the plasma LDL-cholesterol concentration of approximately 10–15 mg/dl would be realized. Since the majority of patients with elevated levels of plasma LDL-cholesterol have overproduction as one component of their hypercholesterolemia, this is an exceedingly important avenue for future pharmacological developments.

The second major factor that governs the final concentration of plasma LDL cholesterol is the rate of receptor-dependent LDL uptake. From work done in the experimental animal it has become apparent that approximately 90% of all receptor-dependent transport manifest in the whole animal is localized in the liver and, to a lesser extent, the small intestine. It is likely that the same situation is true in man [18]. Since these are also the two major organs responsible for the excretion of cholesterol and its major metabolites from the body, it follows that the second major approach to the reduction of plasma LDL-cholesterol levels is to increase LDL receptor activity in these two organs. It has already been shown that by limiting the availability of cholesterol to the cell, it is possible to increase receptor-dependent transport into that tissue [7,8,19]. In the intestine, this might be accomplished by a combination of agents that both block cholesterol uptake into the epithelial cell from the intestinal lumen and partially inhibit the adaptive increase in cholesterol synthesis that occurs under these circumstances [4]. Alternatively, cholesterol loss from the liver could be enhanced with agents such as bile acid sequesterants while, at the same time, inhibiting the adaptive increase in hepatic cholesterol synthesis. In both of these situations the increase in receptor-dependent LDL uptake from the plasma would significantly lower the circulating levels of LDL-cholesterol. In man, for example, doubling the amount of receptor-dependent LDL transport would essentially halve the circulating LDL-cholesterol concentration (Fig. 8).

Thus, in summary, high levels of LDL-cholesterol in the plasma are a major risk factor in the genesis of atherosclerosis, a condition that is directly linked to a variety of common and serious human diseases. Since reducing the plasma LDL-cholesterol concentration has been shown to decrease the risk of these diseases, there is now a concerted effort underway to develop safe and effective methods for accomplishing this. In the past, progress in this area has been hampered by a lack of knowledge concerning the physiology of LDL metabolism. However, this situation is now changing and techniques are presently available to actually quantitate the factors that dictate plasma LDL-cholesterol concentrations in the experimental animal and in man. Thus, for the first time, it is possible to target new drug designs at those areas of LDL metabolism that have the greatest impact on the plasma LDL-cholesterol concentration. From these considerations it is apparent that there are four parameters that modulate the plasma LDL-cholesterol concentration: these include the LDL-cholesterol production rate (J_t), the amount of receptor-dependent LDL transport evident in the body (J_m), the affinity of the LDL receptors for the LDL molecule (K_m) and the rate of receptor-independent uptake (P). Of these four parameters, however, only two, the LDL-cholesterol production rate and the LDL receptor number, have quantitatively important effects upon the plasma LDL-cholesterol concentration. Furthermore, since the majority of people with high circulating levels of LDL-cholesterol appear to have some degree of overproduction, it is reasonable to direct new forms of therapy at this step as well as at increasing receptor activity.

7 Acknowledgements

This work was supported by U. S. Public Health Service Research grants HL-09610 and AM-19329 and by a grant from the Moss Heart Fund. Dr. Meddings was supported by a grant from the Alberta Heritage Foundation.

8 References

1. Rudel, L. L., Parks, J. S., Johnson, F. L., Babiak, J. J.: Lipid Res. 27, 465 (1986)
2. Lipid Research Clinics Program JAMA 251, 351 (1984)
3. Westergaard, H., Dietschy, J. M.: The uptake of lipids into the intestinal mucosa, in: Physiology of Membrane Disorders (eds.) Andreoli, T. E., Hoffman, J. F., Fanestil, D. D., Schultz, S. G., p. 597, New York, Plenum Publishing Corporation 1986.

4. Stange, E. F., Dietschy, J. M.: Cholesterol absorption and metabolism by the intestinal epithelium, in: Sterols and Bile Acids (eds.) Danielsson, H., Sjovall, J., p. 121, Amsterdam, Elsevier Science Publishers B. V. 1985
5. Sherrill, B. C., Dietschy, J. M.: J. Biol. Chem. 253, 1859 (1978)
6. Yamada, N., Shames, D. M., Stoudemire, J. B., Havel, R. J.: Proc. Natl. Acad. Sci. USA 83, 3479 (1986)
7. Spady, D. K., Bilheimer, D. W., Dietschy, J. M.: Proc. Natl. Acad. Sci. USA 80, 3499 (1983)
8. Brown, M. S., Goldstein, J. L.: Science 232, 34 (1986)
9. Spady, D. K., Turley, S. D., Dietschy, J. M.: J. Clin. Invest. 76, 1113 (1985)
10. Spady, D. K., Huettinger, M., Bilheimer, D. W., Dietschy, J. M.: J. Lipid Res. 28, 32 (1987)
11. Dietschy, J. M.: General principles governing movement of lipids across biological membranes, in: Disturbances in Lipid and Lipoprotein Metabolism (eds.) Dietschy, J. M., Gotto, A. M., Ontko, J. A., p. 1, Baltimore, Waverly Press, Inc. 1978.
12. Spady, D. K., Meddings, J. B., Dietschy, J. M.: J. Clin. Invest. 77, 1474 (1986)
13. Meddings, J. B., Dietschy, J. M.: Circulation 74, 805 (1986)
14. Bilheimer, D. W., Stone, N. J., Grundy, S. M.: J. Clin. Invest. 64, 524 (1979)
15. Grundy, S. M., Vega, G. L.: J. Lipid Res. 16, 1464 (1985)
16. Bilheimer, D. W., Grundy, S. M., Brown, M. S., Goldstein, J. L.: Proc. Natl. Acad. Sci. USA 80: 4124 (1983)
17. Spady, D. K., Dietschy, J. M.: Proc. Natl. Acad. Sci. USA 82, 4526 (1985)
18. Bilheimer, D. W., Goldstein, J. L., Grundy, S. M., Starzl, T. E., Brown, M. S.: N. Engl. J. Med. 311, 1658 (1984)
19. Kovanen, P. T., Bilheimer, D. W., Goldstein, J. L., Haramillo, J. J., Brown, M. S.: Proc. Natl. Acad. Sci. USA 78, 1194 (1981)

Biopharmacological Regulation of Protein Phosphorylation

Hiroyoshi Hidaka and Masatoshi Hagiwara

Department of Molecular and Cellular Pharmacology,
Mie University School of Medicine, Edobashi, Tsu, Mie 514, Japan

Biochemical approaches are, so-called, one way traffic and it is often difficult to prove that the reaction evidenced in vitro is functional in vivo. In such cases, techniques involving molecular pharmacological approaches using selective inhibitors or activators come to the rescue like two way traffic. This article indicates pharmacological approaches to the roles which multiple protein kinases play in regulating various cell functions, using selective inhibitors or activators of cyclic nucleotide-dependent protein kinase, Ca^{2+}-CaM-dependent MLC-kinase and protein kinase C. In this chapter, we indicate the inhibition or activation mechanisms of each compound and examine selectivity for each kinase, then describe the representative experiments of biological applications of these compounds. The possible clinical application of these compounds warrant attention.

Progress in Clinical Biochemistry
and Medicine, Vol. 5

1 Introduction

Cyclic nucleotides, Ca^{2+} ions, and diacylglycerol are important intracellular messengers. Until the 1970's, the pharmacological approaches focused on changes in functions of the cell, in response of intracellular Ca^{2+} and cyclic nucleotide concentration, and/or on mobilization using Ca^{2+} channel blockers, cyclic nucleotide phosphodiesterase inhibitors, and adenylate or guanylate cyclase activators [1-3]. These experiments led to the conclusion that a rise in intracellular cyclic nucleotides concentration results in binding to a single small class of receptor proteins which interact with a single class of response elements, protein kinases components, which catalyze a single type of reaction, that is protein phosphorylation [4]. Pharmacological studies using the calmodulin (CaM) antagonists, phenothiazines and naphthalenesulfonamides suggested that Ca^{2+}-CaM-dependent phosphorylation may play an important role in functions in various tissues [5]. In 1979, protein kinase C was found to be a Ca^{2+}- and phospholipid-dependent enzyme activated by diacylglycerol [6]. Hence, the biological functions of cyclic nucleotides, Ca^{2+}-CaM, and diacyl glycerol seem to be manifest through protein phosphorylation by cyclic nucleotides, Ca^{2+}-CaM-dependent protein kinases, and protein kinase C [7,8].

Pharmacological agents can be used to determine the physiological significance of protein phosphorylation in biological systems. However, to elucidate the physiological function of each protein kinase, these agents must meet definite criteria; 1) direct binding to the protein kinase, 2) strict specificity for a protein kinase, 3) cell membrane permeable, 4) rapidly synthesizing considerable amounts of the agent. A number of natural or synthesized modulators of protein kinases have been prepared (Table 1), yet only several compounds of the isoquinolinesulfonamides and naphthalenesulfonamides, meet the criteria. For example most of the protein kinase C inhibitors listed in Table 1 are lipophilic and interact with phospholipid, hence there is an indirect inhibition of protein kinase C. Moreover, these inhibitory agents are relatively nonselective in their action on phospholipid/Ca^{2+}- and calmodulin/Ca^{2+}-dependent enzymes. In this chapter, the question addressed is the role which multiple protein kinases play in regulating various cell functions and attention is directed to available novel compounds which have a high selectivity for cyclic nucleotide-dependent protein kinases, Ca^{2+}-CaM-dependent MLC-kinase, and protein kinase C.

2 Selective Inhibitors and Activators of Protein Kinases

2.1 Myosin Light Chain Kinase (MLCK)

Intracellular calcium is a second messenger, the diffusion of which enables a signal to propagate rapidly throughout the cell. Calcium binds to the intracellular Ca^{2+} receptor, calmodulin and/or to other Ca^{2+} binding proteins. The second messenger Ca^{2+} acts through Ca^{2+} binding proteins by activating the enzyme, protein kinase, which then phosphorylates a protein. With phosphorylation, the protein changes shape.

In 1977, the calmodulin antagonist, naphthalenesulfonamide (W-7) which we synthesized was first introduced at the USA-Japan Joint Congress in Hawaii. W-7 relaxes vascular smooth muscle and inhibits superprecipitation of actomyosin from aortic smooth muscle, thereby indicating a role for calmodulin in muscle contraction [5]. This hypothesis was given support in two different types of experiments, reported by Dabrowska *et al.* [9] and Hidaka *et al.* [10]. We propose that calmodulin antagonists be defined as agents which bind to CaM, Ca^{2+} dependently, and inhibit selectively the Ca^{2+}-CaM-dependent enzyme. Since Honda *et al.* [11] first suggested that phenothiazines may be calmodulin antagonists, a large number of CaM antagonists appeared. During the synthesizing and selecting of a novel CaM inhibitor among the derivatives of naphthalenesulfonamide, we discovered that shorter alkyl chain derivatives of W-7 (N-(6-aminohexyl)-5-chloro-1-naphthalenesulfonamide) markedly inhibit protein kinases such as myosin light chain kinase, cyclic AMP dependent protein kinase and cyclic GMP dependent protein kinase, through mechanisms different from these of W-7. Alternatively, a shorter alkyl chain derivative of W-7, N-(6-aminoethyl)-5-chloro-1-naphthalenesulfonamide (A-3) directly inhibited these protein kinases without interaction with CaM [12,13]. Finally, one of these derivatives, 1-(5-chloronaphthalene-1-sulfonyl)-1 H-hexahydro-1,4-diazepine (ML-9) proved to be a specific inhibitor of MLCK [14]. Kinetic analysis indicated that ML-9 inhibited MLC-kinase competitively with respect to ATP but not myosin light chain or CaM, and that the K_i value of ML-9 against MLCK was 3.8 μM. Accordingly this compound appears to interact with MLCK at the binding site or near the site of ATP. ML-9 was then used as a pertinent pharmacological tool for studying physiological roles of MLCK. In various tissues and even platelets, ML-9 proved to be useful for studying the role of MLCK-induced phosphorylation in platelet activation phenomena.

2.2 Protein Kinase C

At higher concentrations, W-7 inhibits Ca^{2+}-phospholipid-dependent protein kinase (protein kinase C) activity, in a manner competitive with that of phospholipid [13].

When the naphthalene ring of the naphthalenesulfonamides was replaced by isoquinoline, the derivatives were no longer CaM or phospholipid interacting agents but rather compounds which directly suppress protein kinases. Among them, 1-(5-isoquinoline-sulfonyl)-2-methylpiperazine, referred to as H-7 is a relatively selective inhibitor of protein kinase C[15]. The K_i value of this compound against protein kinase C is 6.0 μM. Kinetic analysis by double reciprocal plots revealed that inhibition of protein kinase C by H-7 was competitive to ATP and noncompetitive with respect to the phosphate acceptor. Thus, it was clearly shown that H-7 directly inhibits protein kinase C, and not through interaction with Ca^{2+} or phospholipid. H-7 is now in wide use for studies on various biological systems. It has significant effects on various functions by inhibiting protein kinase C-induced phosphorylation in cells [16-31].

On the other hand, a derivative of W-7 (N-(6-phenylhexyl)-5-chloro-1-naphthalenesulfonamide (SC-9) is a Ca^{2+}-dependent activator of protein kinase C [32]. SC-9 and phosphatidylserine have similar effects on protein kinase C activity, as follows. 1) SC-9 also activates protein kinase C in a Ca^{2+} dependent manner with a similar Km value for Ca^{2+}. 2) TPA also reduces the requirement for Ca^{2+} of the SC-9 activated protein kinase C and the apparent Km value for TPA of SC-9-activated protein kinase C is similar to that of the phosphatidylserine-activated enzyme. 3) Al-

Table 1.

	Protein kinase C		MLC-kinase		cAMP-dependent protein kinase	
Direct Inhibitors	Isoquinolinesulfonamides		Naphthalenesulfonamide		Isoquinolinesulfonamides	
	H-7	*Hidaka et al. (1984)	A-3	*Inagaki et al. (1986)	H-8	*Hidaka et al. (1984)
	H-9	*Hidaka et al. (1984)	ML-9	*Saitoh et al. (1986)	ATP analogs	Hoppe et al. (1978)
Indirect Inhibitors	Anticalmodulins		Anticalmodulins		Affinity labeling drugs	
	Phenothiazines	Wise & Kuo (1983)	Phenothiazines	Weiss & Levin (1978)	5′-FSU_2BzAdo	Hixon & Krebs (1979)
	W-7	*Tanaka et al. (1982)	W-7	*Hidaka et al. (1979)	Cibacrone Blue	Witt & Roskouski (1980)
	Calmidazolium	Mazzei et al. (1984)	Calmidazolium	Mazzei et al. (1984)	DTNB	Armstrong et al. (1978)
	Polypeptides		Polypeptides		NBD-Cl	Hartl et al. (1982)
	Cytotoxins	Kuo et al. (1983)	Cytotoxins	Katoh et al. (1982)	o-Phthaladehyde	Puri et al. (1985)
	Polymixin B	Mazzei et al. (1984)	Neurotoxins	Mazzei et al. (1982)	Polypeptides	
	Neurotoxins	Mazzei et al. (1984)			Protein kinase inhibitor	Walsh et al. (1971)
	Polyamines					
	Spermine	Qi et al. (1984)				
	1,12-Diaminododecane	Qi et al. (1984)				
	Local Anesthetics					
	Dibucaines	Mori et al. (1980)				
	Others					
	Adriamycin	Wise & Kuo (1983)				
	Palmitoylcarnitine	Wise & Kuo (1983)			cAMP analogs	Haaster et al. (1984)
	Alkyllysophospholipid	Helfman et al. (1983)				
	Gangliosides	Kim (1986)				
	Sphingosine	Hannun et al. (1986)				
	Staurosporin	Tamaoki et al. (1986)				
	Quercetin	Gischwendt et al. (1983)				
	Neomycin	*Hagiwara et al. (unpublished)				

Activators	Phorbolesters		cAMP analogs	Corbin et al. (1985)
	TPA	Castagna et al. (1982)		
	PDBu	Castagna et al. (1982)		
	Diacylglycerol			
	Diolein	Takai et al. (1979)		
	OAG	Mori et al. (1982)		
	Teleocidin	Fujiki et al. (1981)		
	Aplysiatoxin	Fujiki (1981)		
	Saccharin	Kleine (1986)		
	Dexamethasone	Kleine (1986)		
	SC-9	*Ito et al. (1986)		

* From our laboratory

though SC-9 produced further stimulation of myosin light chain phosphorylation by protein kinase C with a low concentration of phosphatidylserine, no further enhancement of the enzymatic activity was observed in the presence of a saturated amount of phosphatidylserine. 4) The calmodulin antagonist W-7, inhibits the phosphatidylserine-induced protein kinase C activity competitively with respect to phosphatidylserine, and also the myosin light chain phosphorylation catalyzed by SC-9-activated protein kinase C.

Pharmacological activators of protein kinase C, such as SC-9, will aid in elucidating the biological role of phosphorylation of this protein in intact cells. Exposure of Swiss 3T3 cells to SC-9 led to a dose related increase in hexose uptake, an event also observed when the cells were treated with TPA. This activation by SC-9 was inhibited by H-7, a specific inhibitor of protein kinase C [72]. If SC-9 activates protein kinase C in intact cells (as expected), in vivo studies will be facilitated.

2.3 Cyclic Nucleotide Dependent Protein Kinase

Since the discovery of cAMP-dependent protein kinase [33] and the evidence that this kinase mediates hormonal stimuli leading to a breakdown in glycogen, enzymatic phosphorylation and dephosphorylation, attention has been directed to the possible molecular mechanisms of biological regulation by which intracellular events are controlled by external stimuli.

We showed that inhibitors of cyclic nucleotide phosphodiesterase (PDE) hydrolysis of cyclic AMP and cyclic GMP affect a variety of physiological responses such as platelet aggregation [34–36], vascular relaxation [37, 38], secretory exocytosis [39] and cell proliferation [40, 41]. Inhibitors of cyclic nucleotide PDE are often used as pharmacological tools to demonstrate that cyclic nucleotides play important roles as second messengers, in various physiological events.

We recently found that derivatives of W-7 exhibit a selective inhibition toward cyclic nucleotide-dependent protein kinases and that N-[2-(methylamino)ethyl]-5-isoquinolinesulfonamide (H-8) was the most potent cyclic nucleotide-dependent protein kinase inhibitor in this series of compounds. The inhibition was freely reversible and of the competitive type, with respect to ATP [15]. We examined the interaction of H-8 with protein kinase using various affinity labeling reagents of the ATP binding site of the purified protein kinase, catalytic subunit and the gel-permeation binding assay of Hummel and Dreyer [42]. We found that H-8 specifically binds to the ATP binding site of the catalytic subunit with binding ratio of 1:1 (Fig. 1) and that H-8 has unique features which differ from the ATP analogues reported by Hoppe et al. [43] or others, in the following points. a) H-8, among other protein kinases or ATP utilizing enzymes, specifically inhibits cyclic nucleotide-dependent protein kinase. b) The binding constant (Km) of H-8 to the enzyme was much lower than that of ATP. c) The binding of H-8 to the enzyme was independent of magnesium ion. d) The binding subsite of H-8 at the active site of the enzyme slightly differed from that of the ATP (Fig. 2) [44].

The structures and characteristics of the aforementioned are summarized in Fig. 3.

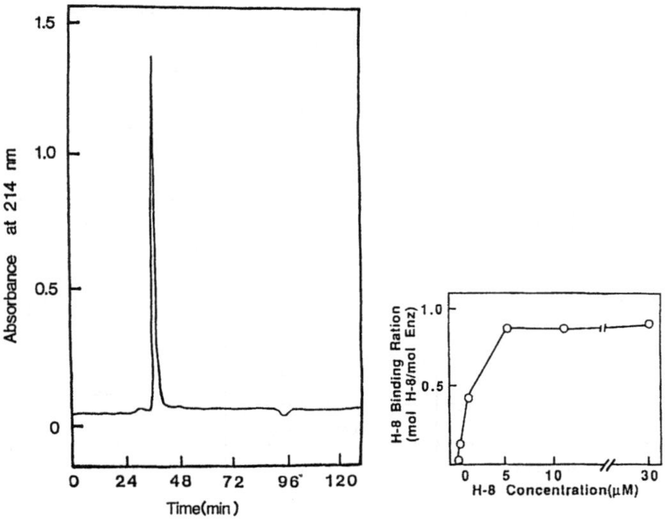

Fig. 1. Binding of H-8 to the catalytic subunit of cAMP-dependent protein kinase, determined by high-performance gel filtration liquid chromatography. Measurements were done according to Hummel & Dreyer [44]. The samples were applied to a column (0.6 × 30 cm) of TSK GEL G3000 SW equilibrated in 50 mM potasium phosphate, pH 6.8, 150 mM NaCl and 1 mM N-[2-(methylamino)-ethyl]-5-isoquinolinesulfonamide (H-8) at 0.5 ml/min. (A) An aliquot (25 μl) of the catalytic subunit solution (10 μM) equilibrated in the column buffer was applied to the column and the elution was monitored with absorbance at 214 nm. The inset is the equilibrium binding of H-8 by the catalytic subunit of cAMP-dependent protein kinase as a function of H-8 concentration. The ordinate expresses the number of mol of H-8 bound per mol of protein. The amount of enzyme molecule was calculated considering the activity lost through penetration of the column. The abscissa shows the concentration of H-8. Results are the means of values calculated from external and internal calibrations, and the range of values was less than 10%.

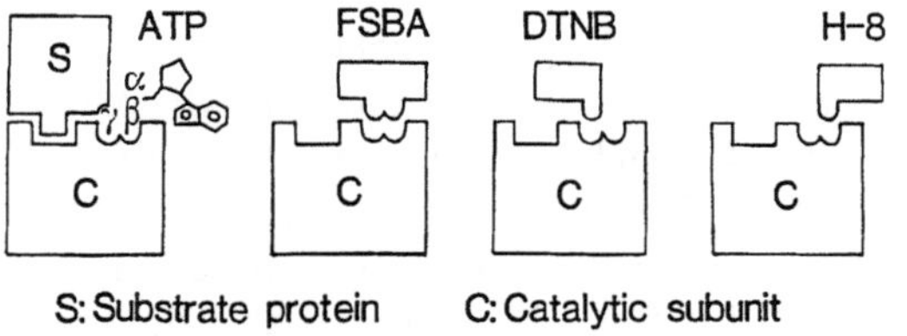

Fig. 2. A model of the H-8 binding site of the catalytic subunit

3 Application of Selective Inhibitors for Affinity Chromatography

Protein kinase C from bovine heart [45], pig spleen [46], and rat brain [47,48] has been purified using DEAE-cellulose, gel filtration and hydrophobic chromatography. However, these steps of purification require complex and adroit manipulation and the yield is less than 5%. Isoquinolinesulfonamides inhibit several protein kinases, to different extents. Among these compounds, H-9 possesses a free amino group and can be used as a ligand of affinity chromatography. Immobilized H-9 was prepared by direct coupling to Sepharose, using the cyanogen bromide method. The amount of bound H-9 was 10–12 μmol/ml wet polymer. The K_i value of this compound against

	H-7	H-8	ML-9
Structure	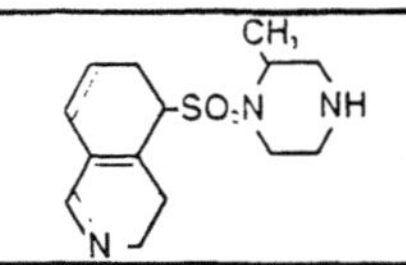	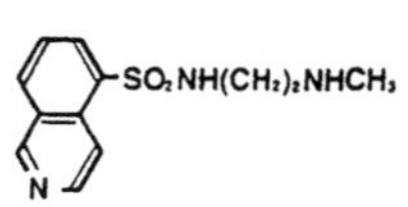	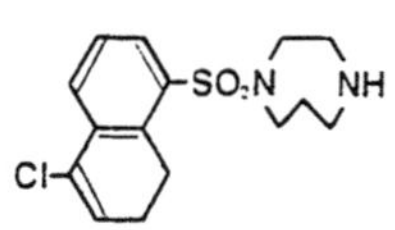
Mode of inhibition	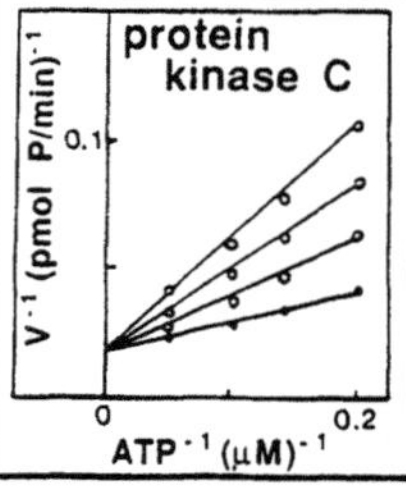	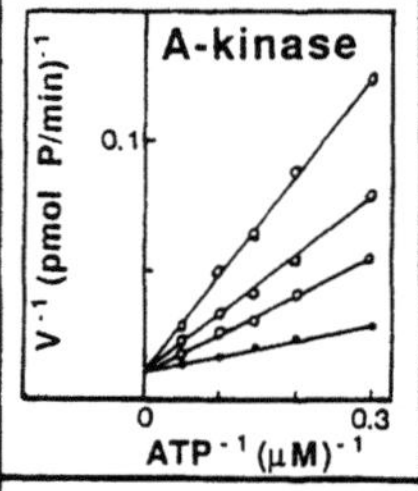	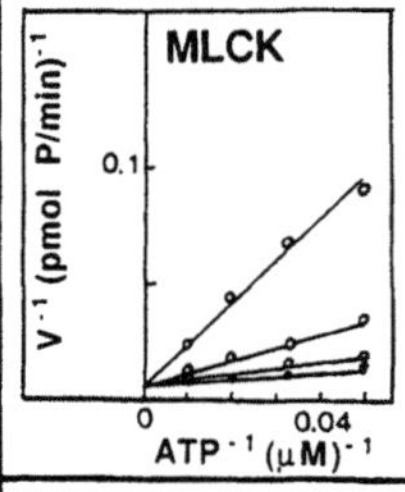
Enzymes	**Ki(μM)**	**Ki(μM)**	**Ki(μM)**
cGMP-dependent protein kinase	5.80	0.48	–
cAMP-dependent protein kinase	3.00	1.20	32
Protein kinase C Ca^{2+}-PS	6.0	14.4	54
Trypsin treated + EGTA	6.0	13.0	–
MLC-kinase Ca^{2+}-CaM	97.0	68.0	3.8
Trypsin treated - EGTA	97.0	65.0	4.0
Casein kinase-I	100	>100	–
Casein kinase-II	>100	>100	–
Actomyosin ATPase (SK)	>100	>100	–
Ca^{2+}-Mg^{2+}ATPase (RBC)	>100	>100	–

Fig. 3. Comparison of protein kinase inhibitors; H-7, H-8 and ML-9

protein kinase C was 18 μM [15]. Three rabbit brains were homogenized in a Potter-Elvenhjem Teflon-glass homogenizer with 285 ml of 25 mM Tris-HCl (pH 7.5) containing 2 mM EGTA, 5 mM DTT, 0.01% Leupepsin and 0.25 M sucrose. The homogenate was centrifuged for 90 min at 100,000 × g. The soluble supernatant, used as the crude extract (300 ml, 960 mg of protein), was applied to a DEAE column (∅ 2 × 20 cm) equilibrated with 25 mM Tris-HCl (pH 7.5) containing 2 mM EGTA, 5 mM DTT and 0.002% Leupepsin (Buffer A). The column was washed with 500 ml of the same solution. The enzyme was eluted from the column by application of a 500 ml linear concentration gradient of NaCl (0.0–0.4 M) Buffer A at a flow rate of 35 ml/h. The eluted fraction (56 ml, 94.8 mg of protein) was dialyzed against Buffer A containing 1 M NaCl and the dialyzed fraction was applied to a Butyl-Toyopearl 650 M column (∅ 5 × 0.5 cm), previously equilibrated with Buffer A containing 1 M NaCl at a flow rate of 30 ml/h. The column was washed with 200 ml of the same buffer at a flow rate of 50 ml/h. The enzyme was eluted with Buffer A containing 0.1 M NaCl at a flow rate of 25 ml/h. Butyl-Toyopearl 650 M fraction (20 ml, 6.4 mg of protein) was applied to an H-9 Sepharose column (∅ 0.8 × 5 cm) equilibrated with Buffer A containing 0.1 M NaCl. The column was washed with 20 ml Buffer A containing 1 mM ATP, followed by 300 ml, of Buffer A containing 1 M NaCl. The protein kinase C activity was eluted with Buffer A containing L-arginine, in a 60 ml linear gradient concentration (0.0–1.5 M) at the flow rate of 15 ml/h. Figure 4b shows a typical elution profile of protein kinase C. A typical purification of the protein kinase

Purification of protein kinase C from rabbit brain

Fraction	Volume	Specific activity	Protein	Activity	Purification	Yield
	ml	nmolPi/min/mg	mg	nmolPi/min	-fold	%
Supernatant	300	0.36	960	348.852	1	100
DE52-eluted	56	2.24	94.8	212.716	6.2	61
Butyl-Toyopearl 650M-eluted	20	13.76	6.4	88.060	38.2	25
H-9 affinity Sepharose 4B-eluted	10	271.1	0.1	37.105	753	8

a

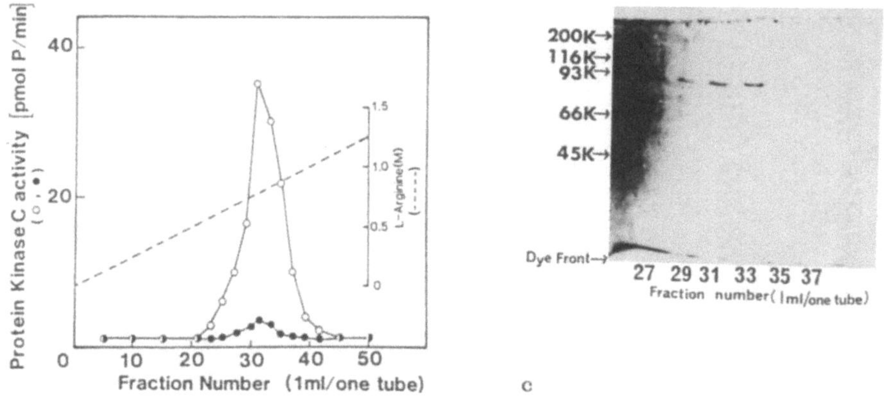

b c

Fig. 4. a: Summary of protein kinase C purification. b: Elution profile of protein kinase from H-9 affinity Sepharose 4B-eluted (○: with Ca^{2+}-PS, ●: with EGTA). c: SDS-PAGE of eluted fractions in Fig. 4b.

is summarized in Table. The enzyme was purified 753-fold compared to the 100,000 × g supernatant. The purity and homogeneity were determined by polyacrylamide gel electrophoresis in the presence of sodium dodecyl sulfate. The elution profile of one major band of Mr 84,000 was found to correspond to that of protein kinase C activity (Figs. 4b and c). We, therefore, attempted to purify protein kinase C using an H-9-Sepharose gel. This method using H-9 affinity chromatography facilitates a large scale preparation of the enzyme for peptide sequencing [49].

The purified enzyme which gave a single band upon SDS polyacrylamide gel electrophoresis was digested with lysylendopeptidase. The peptides were separated by reverse phase HPLC and four peptide sequences were determined by a gas-phase sequencer. Three sets of oligonucleotide mixtures were synthesized with a DNA synthesizer.

A rabbit cDNA library was screened using synthetic oligonucleotide mixtures or fragments of cloned cDNA as probes. Isolated positive clones were classified into three types, α, β and γ, on the basis of restriction mapping, nucleotide sequencing and Southern hybridization.

A comparison of amino acid sequences between protein kinase C types α, β and γ is shown in Fig. 5. Three types of protein kinase C were aligned. Residues in the β or γ type sequences identical to those in the α type are represented by dots. Insertions are represented by hyphens. Basic residues (K and R), H and acidic residues (D and E) are represented as +, · and —, respectively, in the top line when two or all of the three sequences matched.

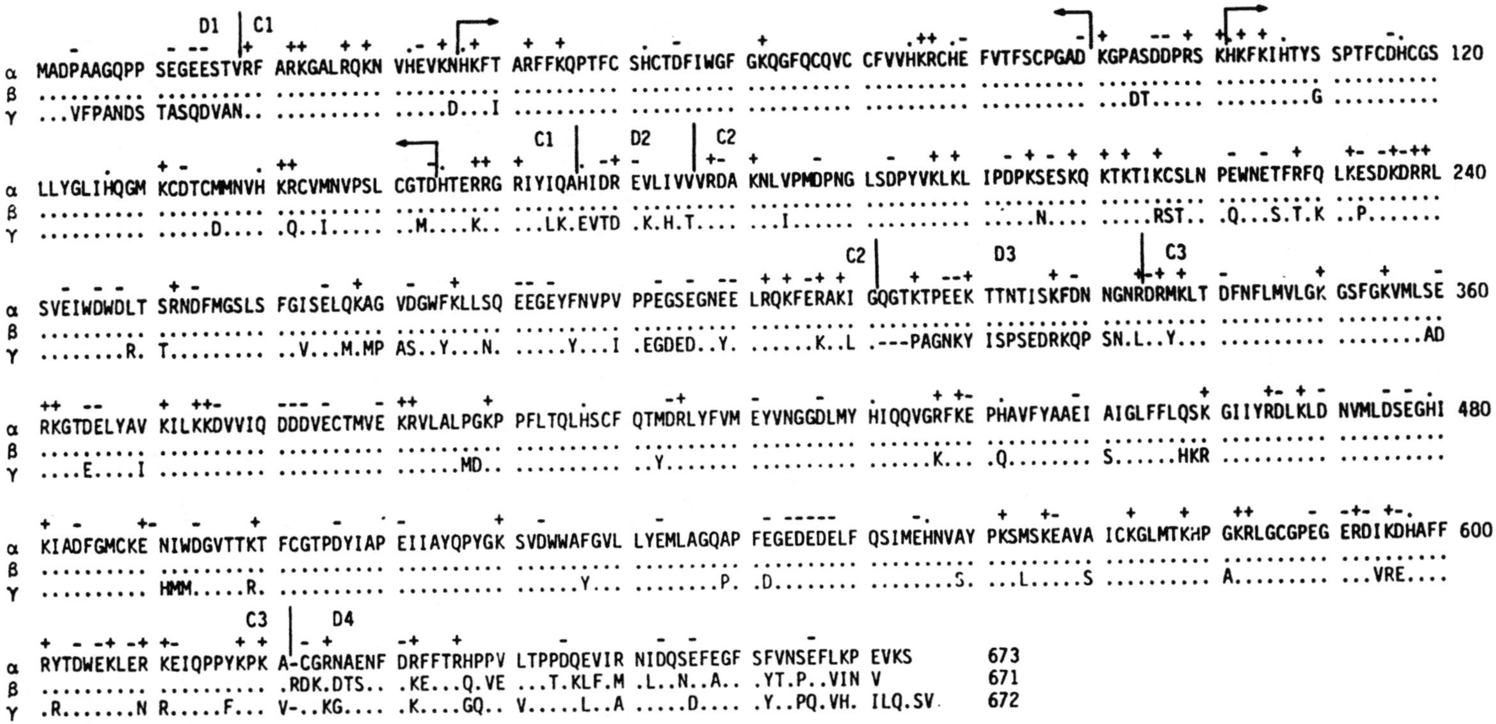

Fig. 5. Amino acid sequence of protein kinase C types α, β and γ.

4 Pharmacological Action of Selective Inhibitors in Vivo

4.1 Platelets

Biological functions of blood platelets including aggregating, secretion, and clot retraction are expressed through cellular contractile activity [50]. The Ca^{2+}-calmodulin-dependent phosphorylation of LC_{20} catalyzed by MLC kinase may be the major regulatory system of the contractile proteins in platelets [51,52], as well as in smooth muscle cells [53]. We found that protein kinase C is responsible for phosphorylation of LC_{20} during platelet activation (particularly in response to stimulation by TPA [54]). Our studies *in vitro* indicated that the phosphorylation by protein kinase C modified the Ca^{2+}-calmodulin-dependent regulation of the actin-myosin interaction [55,56].

H-7 produces significant effects on various cell functions by inhibiting protein kinase C-induced phosphorylation. When H-7 was added to platelets activated by 12-O-tetradecanoylphorbol-13-acetate (TPA) or by thrombin, H-7 enhanced the release of serotonin from human platelets, in association with inhibition of 20,000 dalton protein phosphorylation [17,57].

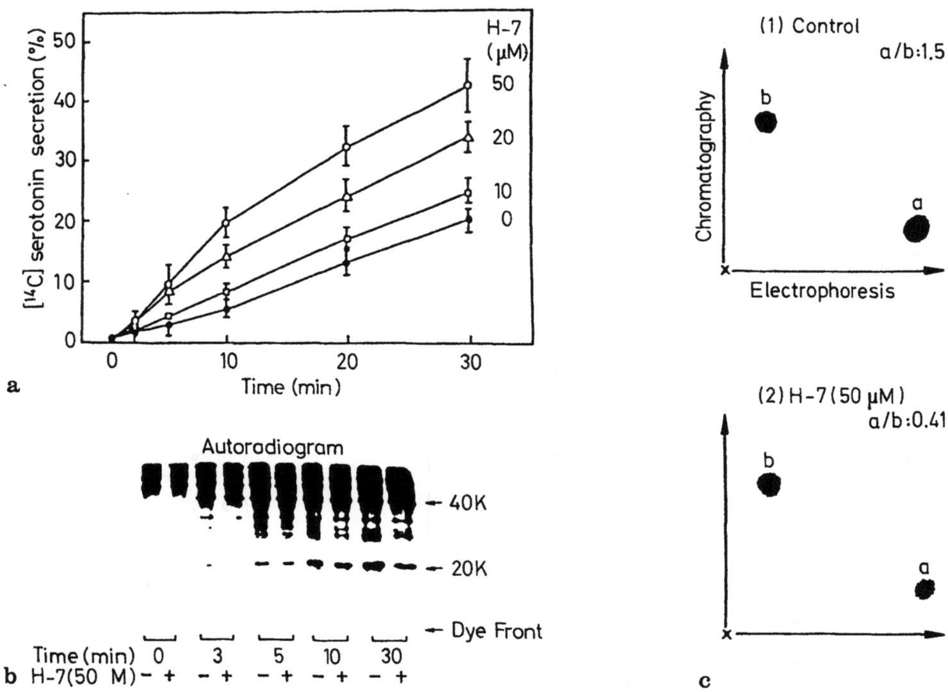

Fig. 6a–c. Time course of ^{14}C-serotonin release (**a**), 20 KDa peptide phosphorylation (**b**), and two-dimensional chromatography of tryptic peptides of the ^{32}P-labeled 20-KDa peptide (**c**) in response to TPA (0.1 μg/ml), and the effect of H-7 preincubation. Serotonin secretion is shown as mean ± S.E. (n = 3) in the absence (●) or presence of H-7, 10 μM (□), 20 μM (△), and 50 μM (○), respectively. Tryptic phosphopeptides following incubation with TPA are indicated in the absence (1) or presence (2) of 50 μM H-7 for 10 min.

H-7 affected neither the protein phosphorylation or the serotonin secretion in unstimulated platelets. A phosphopeptide with a molecular weight of 20,000 has been identified as a light chain (LC_{20}) of platelet myosin and both protein kinase C and Ca^{2+}-calmodulin-dependent myosin light chain kinase are apparently involved in the phosphorylation. Two-dimensional peptide mapping following tryptic hydrolysis revealed that H-7 selectively inhibits the protein kinase C-catalyzed phosphorylation of myosin light chain. This pharmacological evidence suggests that Ca^{2+}-activated, phospholipid-dependent myosin light chain phosphorylation may play an inhibiotory role in the release reaction induced by TPA (Fig. 6) [58].

It was proposed that protein kinase C serves a negative feedback role in a receptor-mediated cell activation [59-64]. In determine whether the suppressive action by TPA of agonist-induced phosphoinositide turnover and $[Ca^{2+}]_i$ mobilization are mediated through activation of protein kinase C, Tohmatsu et al. investigated the effects of H-7 upon these two metabolic events in TPA-pretreated human platelets eposed to thrombin [20]. Their data indicated that H-7 alleviated the TPA-induced inhibition of phosphoinositide metabolism and Ca^{2+} mobilization in thrombin-activated human platelets, thereby providing evidence that the suppressive effects of TPA are mediated by protein kinase C activation. Poll et al. also showed that H-7 prevents TPA induced inhibition of the NaF mediated rise in $[Ca^{2+}]_i$ and TXB_2 generation [31].

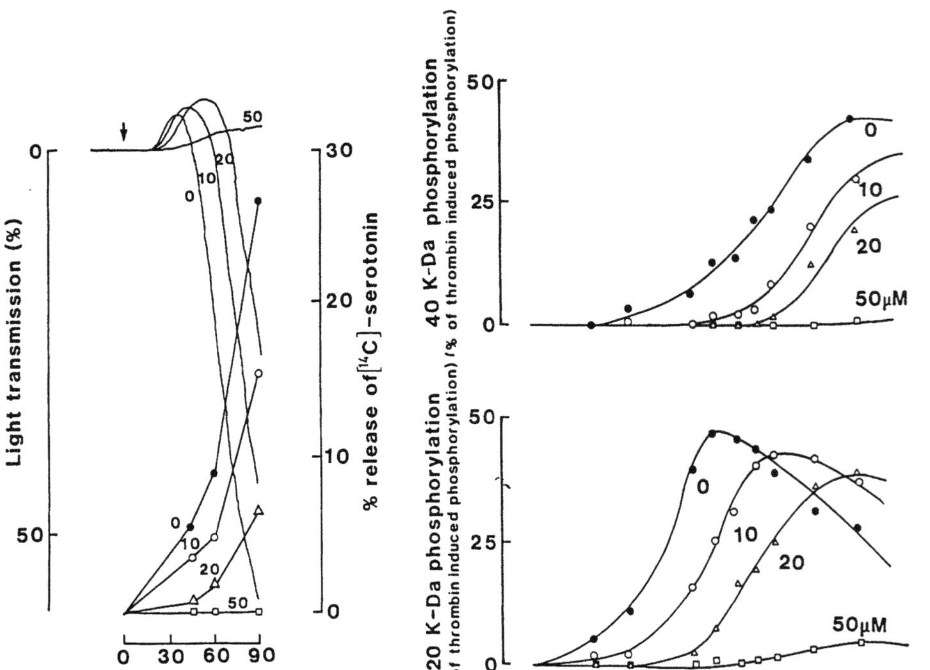

Fig. 7. Time course of platelet aggregation and [^{14}C]-serotonin release (left), and 40 KDa peptide phosphorylation and myosin 20 KDa light chain phosphorylation (right) in response to collagen (1 μg ml^{-1}) and the effect of ML-9 (the final concentration as indicated). Closed circles, control;

When human platelets were stimulated with collagen, myosin 20 K-Da light chain phosphorylation was rapid and appeared to precede change in the shape. Secretion, aggregation and phosphorylation of the 40 K-Da peptide followed. Two-dimensional peptide mapping of this phosphorylated myosin 20 K-Da light chain induced by collagen, following tryptic digestion, revealed that MLC-kinase mainly phosphorylated this myosin 20 K-Da light chain and that the site phosphorylated by protein kinase C was less evident in the early phase during collagen activation (data not shown). ML-9 delayed the time course of the myosin 20 K-Da light chain phosphorylation, in a dose-dependent manner. This delay sequentially led to a delay in aggregation, secretion and phosphorylation of the 40 K-Da peptide, in the same manner (Fig. 7) [14]. There was also a correlation among shape change, myosin 20 K-Da light chain phosphorylation and secretion, as described by Daniel et al. [58]. ML-9 did not affect protein phosphorylation and secretion in the unstimulated platelets. These findings suggest that myosin 20 K-Da light chain phosphorylation by MLC-kinase may be a key event in the activation of platelets with collagen.

4.2 Leucocyte

The effects of TPA on secretion reactions from leucocytes have been extensively investigated the protein kinase C inhibitor, H-7. Okano et al. reported that after mast cells were preincubated in the presence of TPA (100 ng/ml) for 5 min at 37 °C, histamine secretion induced by compound 48/80 (0.5 μg/ml) was repressed to 65% of the maximal response observed without TPA-pretreatment but, in cells treated with H-7, there were no significant changes in the secretory activity [19].

◀

open circles, ML-9 10 μM; open triangles, ML-9 20 μM; open squares, ML-9 50 μM. Methods: Human platelets suspended in Hepes buffered saline (145 mM NaCl, 5 mM KCl, 0.5 mM $MgSO_4$, 10 mM Hepes, 5 mM glucose, pH 7.4 at 37 °C) were incubated for 60 min at 30 °C with either 0.5 mCi^{-1} of [^{32}P] orthophosphate (New England Nuclear) or 0.5 μCi ml^{-1} of [^{14}C]-serotonin (Amersham), spun at 600 xg for 5 min and the supernatant discarded. The pellets were resuspended in the same buffer to give a platelet count of 5–10 × 10^8/ml. Platelet aggregation was studied turbidimetrically using a Rikadenki 4 channel Aggregometer (RAM41). The standard platelet reaction mixtures consisted of 0.45 ml of labelled platelet suspension, 0.05 ml of saline (control) or ML-9 (10, 20 and 50 μM) and collagen (1 μg ml^{-1}). The platelets were preincubated with saline or ML-9 in the aggregometer (Rikadenki) for 1 min before stimulation with collagen. Secretion and protein phosphorylation studies were carried out simultaneously at 37 °C. The reaction was terminated with 0.2 ml of 1.5% formaldehyde for analysis of [^{14}C]-serotonin release, then spun at 1,200 g for 10 min at 4 °C. Samples of supernatant (0.3 ml) were mixed with 10 ml of ACSII (Amersham) and counted in a Beckman scintillation counter. The data are expressed as a percentage of the total platelet [^{14}C]-serotonin measured after lysis of the platelets with 1% Triton X-100. For analysis of phosphoproteins, the reaction was terminated by addition of a half volume of 2-fold-concentrated stock denaturing solution which contained 10% SDS, 6% 2-mercaptoethanol, 20% glycerol, 0.19 M Tris-HCl (pH 6.8). The sample was boiled in a water bath for 3 min and subjected to SDS-polyacrylamide slab gel electrophoresis, as described by Laemmli [12]. The separating and stacking gels contained 15 and 5% acrylamide, respectively. The gel was stained with Coomassie Brilliant Blue, destained, and exposed to a Kodak X-Omat RP X-ray film for autoradiography. The relative intensity of each spot was estimated by densitometric tracing of the autoradiogram using a GS3000 Transmittance/Reflectance Scanning Densitometer. Data for the 40 KDa peptide and myosin 20 KDa light chain were plotted to the responses induced by 0.2 U/ml of thrombin at 40 s, respectively.

Teshima et al., reported that protein kinase C activators such as phosphatidyl serine, phosphatidyl inositol, and phosphatidyl ethanolamine enhanced the phosphorylation of 36 K protein and that H-7 inhibited this phosphorylation [30]. Calcium ions may well regulate a broad spectrum of cellular functions, including proliferation and differentiation. Our groups and Matsui et al. suggested that the calmodulin-dependent process is involved in the proliferation of leucocytes and other cells [65,67]. Matsui et al. recently reported the Ca^{2+} requirement and the protein kinase C involvement in phorbol ester-induced phenotypic differentiation of HL-60 cells, using H-7 as the protein kinase C inhibitor [21]. Wright and Hoffman examined the effect of protein kinase C inhibitors H-7 and H-9 on neutrophil activation [16] and Yamamoto et al. reported that TPA induced interleukin 2 (IL-2) production and IL-2 receptor expression of the human leukemic T-cell line and that this induction was suppressed by H-7 or by W-7. They also observed that HA-1004, a potent inhibitor of cyclic nucleotide-dependent protein kinases, had no effect on IL-2 and IL-2 receptor expression [22]. These findings suggest that protein kinase C and Ca^{2+}-calmodulin are involved in these expressions. Nakao et al. suggested that H-7 inhibits TPA-induced HTLV-I p19 antigen expression in HTLV-I virus-infected human T-cell line KH-2Lo cells [23]. Denis et al. suggested the involvement of protein kinase C in concanavalin A signal transduction of lymphocytes deduced from the finding that H-7 strongly suppressed concanavalin A- and calcium-dependent phosphorylation of a protein Mr 80 KDa in mouse lymphocytes [25]. Ido et al. and our groups showed that H-7 inhibits the TPA-induced reduction of vincristine uptake from P388 murine leukemic cells [27]. The protein kinase C inhibitor H-7 may prove to be a chemotherapeutic drug.

4.3 Insulin Release

Pancreatic islet homogenates have protein kinase C activity and activation of this kinase by either tumor-promoting phorbol esters or the membrane-accessible synthetic diglyceride coincides with the stimulation of insulin release [68]. Protein kinase C conceivably participates in the secretory response to nutrient secretagenous (D-glucose) and cholinergic agents, both of which case hydrolysis of inositol-containing phospholipids in intact islet cells [69], and may increase the availability of endogenous diglyceride. Malaisse and Sener showed that H-7 causes a partial decrease in insulin release evoked by TPA (2.0 μM), at a low concentration of D-glucose (2.8 mM) [24]. H-7 and its derivative failed to affect insulin secretion, at an intermediate glucose concentration (8.3 mM) but decreased insulin output at a higher glucose level (16.7 mM). The increment rate of secretion attributable to carbamylcholine, at the intermediate glucose concentration (8.3 mM), was also decreased in the presence of H-7. These findings suggest that activation of protein kinase C plays a significant role in the functional response of the B-cell to tumor-promoting phorbol esters, cholinergic agents at high concentrations of D-glucose. Pandol and Schoeffield also investigated the function of 1,2-diacylglycerol and protein kinase C on pancreatic enzyme secretion using H-7 [28]. All these biopharmacological effects of H-7 are summarized in Table 2.

Table 2. Effect of protein kinase C inhibitors on cell function

*Insulin release from pancreatic islets (Aichi Gakuin Univ., A. Niki)	Inhibition
*Insulin release from pancreatic islets (Brussel Univ., W. J. Malaisse)	Inhibition
Secretion of pancreatic enzymes (Medical Center California, S. J. Pandol)	Activation
Secretion of pancreatic amylase (Kyoto Univ., H. Adachi)	Inhibition
Histamine secretion from rat peritoneal mast cells (Gifu Univ., Y. Nozawa)	Activation
*Vincristine release from P388 murine leukemic cell (Mie Univ., M. Sakurai)	Inhibition
*Serotonin release from human platelets (Mie Univ., H. Hidaka)	Activation
Release reaction and proliferation of cultured pituitary grand (Tokyo Women's Medical College, E. Omura)	Inhibition
*Serotonin release from rat basophilic leukemia (RBL-2H3) (National Institute of Hygienics, T. Terao)	Inhibition
Rat myoblast (L_6, cell) proliferation (Tokyo Women's Medical College, E. Omura)	Inhibition
Differentiation of HL-60 leukemia cells (Kobe Univ., T. Fujita)	Inhibition
Neutrophils superoxide generation (Kyushu Univ., S. Minakami)	Inhibition
*Superoxide generation in neutrophils (Hiroshima Univ., S. Ishibashi)	Inhibition
PEPCK gene expression (Vanderbilt Univ., D. K. Granner)	Inhibition
Interleukin 2 gene expression (Kumamoto Univ., K. Onoue)	Inhibition
Hexose transport in Swiss 3T3 cells (Kyoto Prefectural Univ., of Med., H. Nishino)	Inhibition
*Concanavalin A signal transduction in lymphocytes (Tokyo Univ., T. Osawa)	Inhibition
HTLV-I-p19 Antigen expression (Kobe Univ., Y. Nakao)	Inhibition
*Promyelocytic leukemia cell HL-60 differentiation (Mie Univ., M. Nishikawa)	Inhibition

* The effect of the inhibitor on protein phosphorylation confirmed

4.4 Sperm Motility

Stimulation of sperm motility by cAMP is mediated by cAMP-dependent protein kinase and phosphorylation of axokinin, a heat stable and NP-40-soluble protein, by cAMP-dependent protein kinase is required for motility, in reactivated sperm models [70]. Tash et al. showed that the cAMP-dependent protein kinase inhibitor, H-8 suppresses motility of viable as well as reactivated sperm, in a dose-dependent manner, and suggested that the phosphorylated axokinin is not only required for flagellar reactivation, but it is sufficient as well [26].

4.5 Vascular Relaxation

We reported that W-7 and its derivatives bind to calmodulin in a Ca^{2+} dependent manner, and with a high affinity. Thus, calmodulin antagonists may induce vascular relaxation by inhibiting the Ca^{2+}-calmodulin activation of MLC-kinase [5]. This hypothesis has gained support in experiments using newly synthesized MLC-kinase inhibitors, A-3 and ML-9. We noted the potent vasodilating effects of these compounds on both skinned and intact vascular strips (unpublished data).

When the naphthalene ring of the compounds, A-3 and ML-9, was replaced by isoquinoline, N-[2-(methylamino)ethyl]-5-isoquinolinesulfonamide (H-8) and N-(2-guanidinoethyl)-5-isoquinolinesulfonamide (HA-1004) were neither calmodulin antagonists nor MLC-kinase inhibitors, rather they potently inhibited cyclic nucleotide-dependent protein kinase, in competition with ATP. H-8 and HA-1004 also relaxed rabbit aortic strips contracted by various agonists and exterted the action at the intracellular or submembranal level [71]. H-8 and HA-1004 are intracellular Ca^{2+} antagonists of afferent class from Ca^{2+} entry blockers such as nicardipine. Excess $CaCl_2$ antagonized the relaxation induced by H-8 and HA-1004 in depolarized smooth muscle, whereas calmodulin antagonist or MLC-kinase inhibitor-induced relaxation was not reversed by excess $CaCl_2$. This suggests that cyclic nucleotide-dependent protein kinases may interact with Ca^{2+}-mobilization, at intracellular levels.

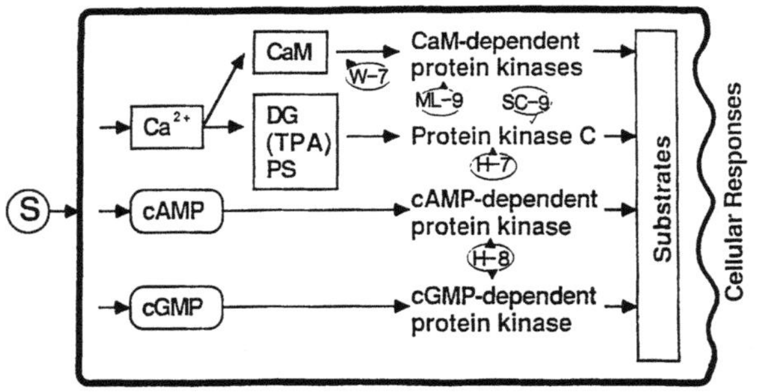

Fig. 8. Pharmacological manipulation of intracellular messenger systems

5 Conclusion

The pharmacological aspects and molecular mechanisms described herein are illustrated in Fig. 8. All the aformentioned evidence suggests that the biochemical function of each protein kinase depends on the tissue in question. While biochemical studies on the functions of protein kinases are feasible, *in vivo* studies are relatively difficult to carry out. Thus a specific inhibitor or activator could pave the way for elucidation of the function of each protein kinase. The possible clinical application of these compounds warrants attention.

6 Acknowledgements

We are indebted to our collaborators for carrying out the experiments presented in this chapter. We also thank Dr. Y. Nagai for kind advice on the presentation of the work and M. Ohara of Kyushu University fo assistance in preparing this article.

This investigation was supported in part by Grants-In-Aid for Scientific Research from the Ministry of Education, Science, and Culture, Japan.

7 References

1. Fleckenstein, A.: Circulation *52*, 1–3 (1983)
2. Hidaka, H., Tanaka, T., Itoh, H.: Trends in Pharmacological Sciences *5*, 237 (1984)
3. Hidaka, H., Endo, T.: Selective inhibitors of three forms of cyclic nucleotide phosphodiesterase–basic and potential clinical applications, in Advances in Cyclic Nucleotide and Protein Phosphorylation Research, Vol. 16 (ed.) Strada, S. J., Thompson, J. W., Raven Press, New York (1984)
4. Rasmussen, H.: Calcium and cAMP as synarchic messengers, New York Wiley (1981)
5. Hidaka, H., Yamaki, T., Asano, M., Totsuka, T.: Blood Vessels *15*, 55 (1978)
6. Takai, Y., Kishimoto, A., Kikkawa, U., Nishizuka, Y.: Biochem. Biophys. Res. Commun. *91*, 1218 (1979)
7. Greengard, P.: Science *199*, 146 (1978)
8. Nishizuka, Y.: Nature *5961*, 693 (1984)
9. Dabrowska, R., Rherry, J. M. F., Aromatorio, D. K., Hartshorne, D. J.: Biochemistry *17*, 253 (1978)
10. Hidaka, H., Yamaki, T., Totsuka, T., Asano, M.: Mol. Pharmacol. *15*, 49 (1979)
11. Honda, F., Katsuki, S., Sakai, N.: Jap. J. Pharmacol. *63*, 530 (1973)
12. Inagaki, M., Kawamoto, H., Itoh, M., Saitoh, M., Hagiwara, M., Takahashi, J., Hidaka, H.: Mol. Pharmacol. *29*, 577 (1986)
13. Tanaka, T., Ohmura, T., Yamakado, T., Hidaka, T.: Mol. Pharmacol. *22*, 408 (1982)
14. Saitoh, M., Naka, M., Hidaka, H.: Biochem. Biophys. Res. Commun. *140*, 280 (1986)
15. Hidaka, H., Inagaki, M., Kawamoto, S., Sasaki, Y.: Biochemistry *23*, 5036 (1984)
16. Wright, C. D., Hofman, M. D.: Biochem. Biophys. Res. Commun. *135*, 749 (1986)
17. Inagaki, M., Kawamoto, S., Hidaka, H.: J. Biol. Chem. *259*, 14321 (1984)
18. Kawamoto, S., Hidaka, H.: Biochem. Biophys. Res. Commun. *125*, 258 (1984)
19. Okano, Y., Takagi, H., Nakashima, S., Tohmatsu, T., Nozawa, Y.: Biochem. Biophys. Res. Commun. *132*, 110 (1985)
20. Tohmatsu, T., Hattori, H., Nagao, S., Ohki, K., Nozawa, Y.: Biochem. Biophys. Res. Commun. 134: 868 (1986)
21. Matsui, T., Nakao, Y., Koizumi, T., Katakami, Y., Fujita, T.: Cancer Research *46*, 583 (1986)
22. Yamamoto, Y., Ohmura, T., Kawakami, K., Onoue, K., Hidaka, H.: J. Biochem. *100*, 333 (1986)
23. Nakao, Y., Matsuda, S., Matsui, T., Koizumi, T., Katakami, Y., Fujita, T., Ito, Y.: Int. J. Cancer *37*, 911 (1986)
24. Malaisse, W. J., Sener, A.: IRCS Medical Science, in press
25. Denis, G. V., Toyoshima, S., Osawa, T.: Biochem. Biophys. Acta *885*, 136 (1986)
26. Tash, J. S., Hidaka, H., Means, A. R.: J. Biol. Chem. *103*, 649 (1986)
27. Ido, M., Asao, T., Sakurai, M., Inagaki, M., Saitoh, M., Hidaka, H.: Leukemia Res. *10*, 1063 (1986)
28. Pandol, S. J., Schoeffield, M. S.: J. Biol. Chem. *261*, 4438 (1986)
29. Koizumi, T., Nakao, Y., Matsui, T., Katakami, Y., Imanaka, M., Fujita, T.: Molecular Immunology, in press
30. Teshima, R., Suzuki, K., Ikebuchi, H., Terao, T.: Mol. Pharmacol. *23*, 279 (1986)
31. Poll, C., Kyrle, P., Westwick, J.: Biochem. Biophys. Res. Commun. *136*, 381 (1986)
32. Ito, M., Tanaka, T., Inagaki, M., Nakanishi, K., Hidaka, H.: Biochemistry *25*, 4179 (1986)
33. Walsh, D. A., Perkins, J. P., Krebs, E. G.: J. Biol. Chem. *243*, 3763 (1968)

34. Asano, T., Ochiai, Y., Hidaka, H.: Mol. Pharmacol. *13*, 400 (1977)
35. Hidaka, H., Hayashi, H., Kohri, H., Kimura, Y., Hosokawa, T., Igawa, T., Saitoh, Y.: J. Pharmacol. Exp. Ther. *211*, 26 (1979)
36. Hagiwara, M., Endo, T., Kanayama, T., Hidaka, H.: J. Pharmacol. Exp. Ther. *228*, 467 (1984)
37. Hidaka, H., Yamaki, H., Totsuka, T., Asano, M.: Mol. Pharmacol. *15*, 49 (1979)
38. Hagiwara, M., Endo, T., Hidaka, H.: Biochem. Pharmacol. *33*, 453 (1984)
39. Niki, H., Niki, A., Hidaka, H.: Biochemical Research *2*, 413 (1981)
40. Hidaka, H., Sasaki, Y., Tanaka, T., Endo, T., Ohno, S., Fujii, Y., Nagata, T.: Proc. Natl. Acad. Sci. USA *78*, 4354 (1981)
41. Chafouleas, J. G., Bolton, W. E., Hidaka, H., Boyd III, A. E., Means, A. R.: Cell *28*, 41 (1982)
42. Hummel, J. P., Dreyer, W. J.: Biochim. Biophys. Acta *63*, 552 (1962)
43. Flockhart, D. A., Freist, W., Hoppe, J., Lincoln, T. M., Corbin, J. D.: Eur. J. Biochem. *140*, 289 (1984)
44. Hagiwara, M., Inagaki, M., Hidaka, H.: Mol. Pharmacol. (in press)
45. Wise, B. C., Raynor, R. L., Kuo, J. F.: J. Biol. Chem. *257*, 8481 (1982)
46. Schatzman, R. C., Raynor, R. L., Fritz, R. B., Kuo, J. F.: Biochem. J. *209*, 435 (1982)
47. Kikkawa, U., Takai, Y., Minakuchi, R., Inohara, S., Nishizuka, Y.: J. Biol. Chem. *257*, 13341 (1982)
48. LaPeauch, C. J., Ballester, R., Rosen, O. M.: Proc. Natl. Acad. Sci. USA *80*, 6858 (1983)
49. Ohno, S., Kawasaki, H., Imajoh, S., Suzuki, K., Inagaki, M., Yokokura, H., Sakoh, T., Hidaka, H.: Nature, in press
50. Pollard, T. D., Fujiwara, K., Handin, R., Weiss, G.: Ann. N. Y. Acad. Sci. *238*, 218 (1977)
51. Adelstein, R. S., Conti, M. A.: Nature *256*, 597 (1975)
52. Nishikawa, M., Tanaka, T., Hidaka, H.: Nature *287*, 863 (1980)
53. Hartshorne, D. J., Siemankowski, R. F.: Annu. Rev. Physiol. *43*, 519 (1981)
54. Naka, M., Nishikawa, M., Adelstein, R. S., Hidaka, H.: Nature *306*, 490 (1983)
55. Endo, T., Naka, M., Hidaka, H.: Biochem. Biophys. Res. Commun. *105*, 922 (1982)
56. Nishikawa, M., Hidaka, H., Adelstein, R. S., Hidaka, H.: J. Biol. Chem. *258*, 14069 (1983)
57. Daniel, J. L., Molish, I. R., Rigmaiden, M., Stewart, G.: J. Biol. Chem. *259*, 9826 (1984)
58. Hidaka, H., Kawamoto, S.: Protein kinase C-mediated phosphorylation in smooth muscle and platelet myosins, in Phospholipids and Cellular Regulation, Vol. 2 (ed.) Kuo, F. J., CRC Press, Florida 1985
59. de Chaffoy de Courcelles, D., Roevens, P., van Belle, H.: FEBS Lett. *173*, 389 (1984)
60. Halenda, S. P., Feinstein, M. B.: Biochem. Biophys. Res. Commun. *124*, 507 (1984)
61. Zavoico, G. B., Halenda, S. P., Sha'afi, R. I., Feinstein, M. B.: Proc. Natl. Acad. Sci. USA *82*, 3859 (1985)
62. Watson, S. P., Lapetina, E. G.: Proc. Natl. Acad. Sci. USA *83*, 2623 (1985)
63. Sagi-Eisenberg, R., Lieman, H., Pecht, I.: Nature *313*, 59 (1985)
64. MacIntyre, D. E., McNicol, A., Drummond, A. H.: FEBS Lett. *180*, 160 (1985)
65. Matsui, T., Nakao, Y., Koizumi, T., Nakagawa, T., Kobayashi, N., Kishihara, M., Fujita, T.: Cancer Res. *45*, 311 (1985)
66. Sasaki, Y., Hidaka, H.: Biochem. Biophys. Acta *714*, 257 (1982)
67. Means, A. R.: Cell *28*, 41 (1982)
68. Malaissee, W. J., Dunlop, E., Mathias, P. C. F., Malaisse-Lagae, F., Sener, A.: Eur. J. Biochem. *149*, 23 (1985)
69. Best, L., Malaisse, W. J.: Endocrinology *115*, 1814 (1984)
70. Tash, J. S., Kakar, S. S., Means, A. R.: Cell *38*, 551 (1984)
71. Ishikawa, T., Inagaki, M., Watanabe, M., Hidaka, H.: J. Pharmacol. Exp. Ther. *235*, 495 (1985)
72. Nishino, H., Kitagawa, K., Iwashima, A., Ito, M., Tanaka, T., Hidaka, H.: Biochem. Biophys. Acta *889*, 236 (1986)

Calcium-Activated Neutral Protease (CANP) and its Biological and Medical Implications

Koichi Suzuki, Shigeo Ohno, Yasufumi Emori, Shinobu Imajoh, and Hiroshi Kawasaki

Department of Molecular Biology, Tokyo Metropolitan Institute of Medical Science, Honkomagome, Bunkyo-ku, Tokyo 113, Japan

Calcium activated neutral protease (CANP) apparently participates in a variety of cellular functions mediated by Ca^{2+}. The primary structure of CANP reveals a novel domain structure comprising domains highly homologous to papain-like cysteine proteinases and calmodulin-like Ca^{2+} binding proteins, and a glycine-rich hydrophobic domain that interacts with the cell membrane and phospholipid. This finding provides new insight into the protease action, activity-regulation, biological and pathological function and molecular evolution of CANP. The primary structure and gene structure of CANP are described together with discussions on CANP from the above viewpoints.

Progress in Clinical Biochemistry
and Medicine, Vol. 5

1 Introduction

*C*alcium *A*ctivated *N*eutral *P*rotease (CANP) is a typical intracellular cysteine proteinase that exists ubiquitously in the cells of higher animals (reviews see) [1-12]. This protease absolutely requires Ca^{2+} for activity and is optimally active in the neutral pH region. CANP was first found in extracts of rat brain in 1964 [13], but it is only quite recently, with the recognition of the important role of Ca^{2+} in various cellular events, that this protease has received a great deal of attention. There are two isozymes of CANP, μCANP (or calpain I) and mCANP (calpain II). Their various properties are quite similar but they differ clearly in their Ca^{2+} sensitivities [14,15]. While μCANP is active at μM Ca^{2+}, mCANP requires mM Ca^{2+} for its activity. Both isozymes exist in most cells and mCANP is the predominant molecular species. Although CANP is assumed to participate in various cellular events mediated by Ca^{2+}, its biological function is not yet clear.

The amino acid sequence of CANP, determined quite recently by cDNA cloning, reveals that CANP has a clear domain structure comprising a cysteine proteinase domain and a calmodulin-like Ca^{2+} binding domain [16]. This finding significantly stimulated studies on CANP and provided new insights into the protease activity, Ca^{2+} sensitivity, and molecular evolution of CANP. In this review, the structure and function of CANP are discussed mainly on the basis of results obtained since 1984 to avoid overlap with previous reviews [1-12]. Papers published before 1984 are not included in the reference list except for papers of particular importance.

2 Structure of CANP

2.1 Subunit Structure

Both m and μCANPs are heterodimers composed of a large or catalytic subunit (Mr: 80,000, 80 K) and a small or regulatory subunit (Mr: 30,000, 30 K) [6,10,11]. In some cases, a "monomeric" CANP devoid of the 30 K subunit has been reported [6,10,11], but "monomeric" and "dimeric" CANPs are indistinguishable in their various catalytic properties. The partial degradation of the 30 K subunit often takes place during purification, and the nicked 30 K subunit does not usually give a clear band upon SDS polyacrylamide gel electrophoresis. This may be the reason for the apparent "monomeric" nature of CANP. The structure of the chicken 80 K subunit was first deduced in 1984 from the nucleotide sequence of complementary DNA (cDNA) [16]. Since then structures of the 80 K and 30 K subunits of rabbit and human CANPs have been determined [17-22]. Sakihama et al. determined the structure of the porcine 30 K subunit independently [23].

Fig. 1. Amino acid sequence of the 80 K subunit of chicken and human CANPs. The identical residues are boxed. The sequence is divided into 4 domains (I–IV) and the protease domain (II) is indicated by a wavy line. The active site amino acid residues, Cys-108 and His-265 (chicken CANP) are shown by triangles. Arabic numbers are the loop regions of EF hand structures ▶

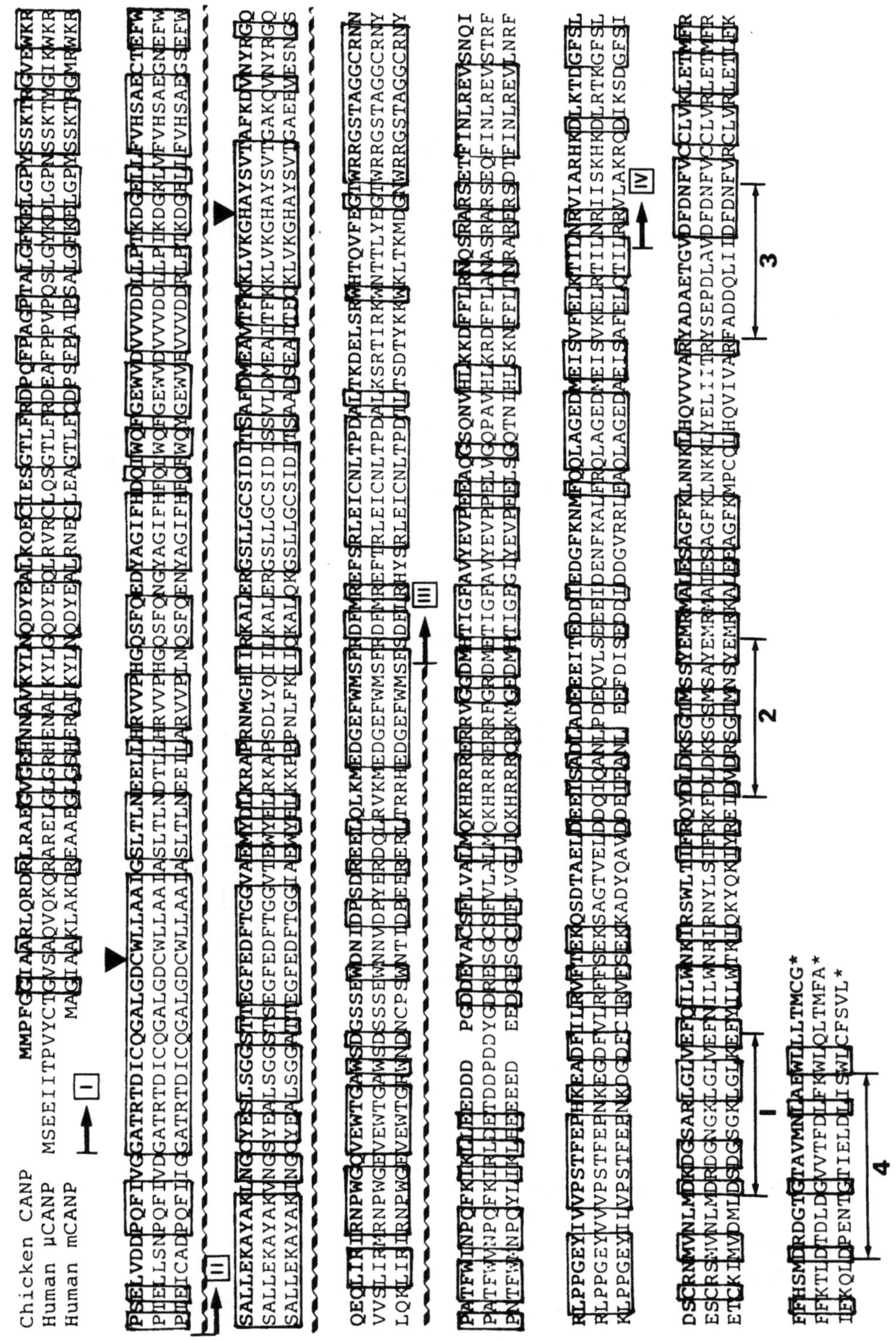
Chicken CANP
Human μCANP
Human mCANP
I
II
III
IV
1
2
3
4

2.2 The Structure of the 80 K Subunit

The structure of the cDNA for chicken CANP was first studied, because only a single molecular species of CANP has been found in chicken and the assignment of cDNA is easier than in mammals where both μ and mCANPs are generally found. Later, the structures of the 80 K subunits of μ and m human and rabbit CANPs were determined similarly by analyzing the cDNA base sequences [17, 19, 21]. The structures of the 80 K subunit of various CANPs are highly homologous with an overall sequence homology of about 60 %. Thus, the 80 K subunits of various CANPs have the same fundamental structure regardless of their origins and Ca^{2+} sensitivities. In the following discussion, the chicken CANP 80 K subunit is taken as a typical example (Fig. 1).

The mRNA for the chicken CANP 80 K subunit is about 3.5 kb long and its primary translation product, composed of 705 amino acid residues, has a molecular weight of 80,350 [16]. The 80 K subunits of human μ and mCANPs consist of 714 (Mr: 81,889) and 700 (Mr: 80,005) amino acid residues, respectively (Fig. 1) [19]. A computer search for sequence homology among proteins of known sequences showed that the 80 K subunit has a clear four-domain structure, I–IV from the N-terminus, as shown in Fig. 2. Domains II and IV are a cysteine proteinase domain and a calmodulin-like Ca^{2+} binding domain, respectively.

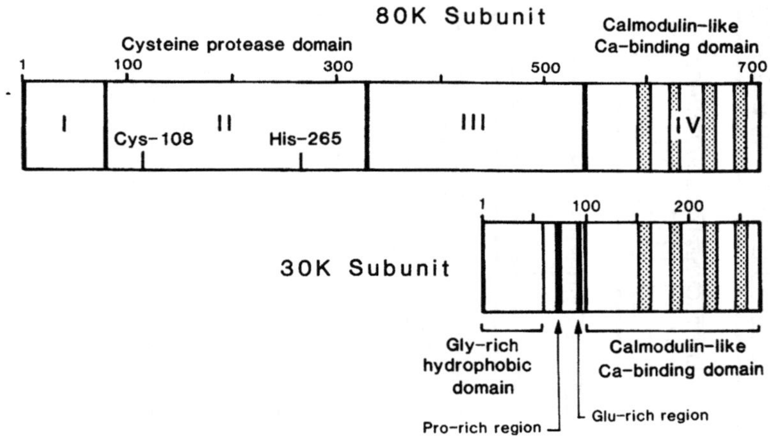

Fig. 2. Domain structure of the 80 K and 30 K subunits of CANP. The 4 striped regions in the C-terminal domain of both subunits indicate EF hand structures

2.2.1 Protease Domain (Domain II, Residues Nos. 81–330)

The structure of this domain is homologous to those of typical cysteine proteinases like cathepsins B and L and papain. The active site cysteine residue in various typical cysteine proteinases is located 25–30 residues from the N-terminus, and the amino acid sequence of this N-terminal region is conserved among cysteine proteinases [24].

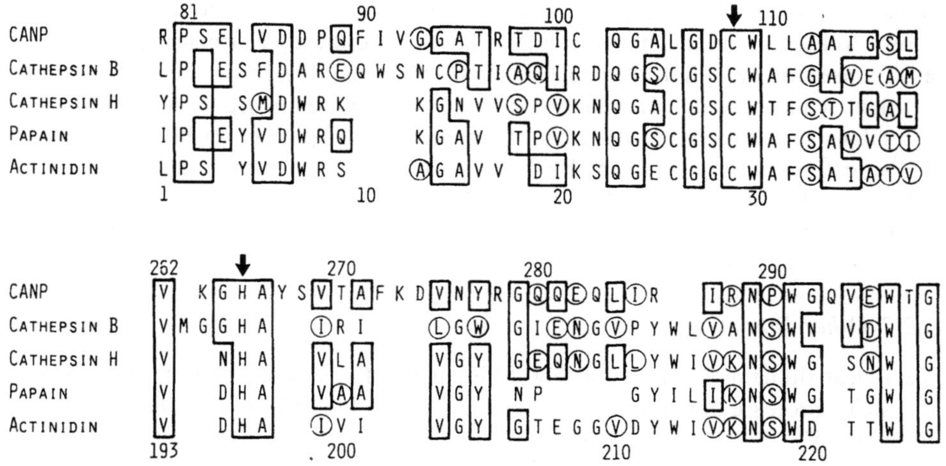

Fig. 3. Comparison of amino acid sequences around the active site cysteine and histidine residues in various cysteine proteinases. Identical amino acids are boxed and favored amino acid substitutions are circled. Numbers above and below the sequences are residue numbers for chicken CANP 80 K subunit and cathepsin B, respectively

The active site of CANP has been identified as Cys-108 [11]. The sequence of Nos. 81–117 in CANP is homologous to this conserved sequence as shown in Fig. 3. The active site Cys-108 is the 28th residue from the assumed N-terminus (Pro-81) of the protease domain [16]. Histidine is another active site amino acid residue in various cysteine proteinases. The amino acid sequence around His-265 in CANP is homologous to that near the active site histidine residue in other cysteine proteinases as shown in Fig. 3 [24]. The involvement of this histidine residue in the catalysis has not been proved, but His-265 presumably constitutes a component of the active site of CANP.

The protease domain in CANP and various cysteine proteinases shows marked sequence homology around two active site amino acid residues but longer sequence homology has not been found in other regions of this protease domain. The protease domain starts at residue No. 81, but the C-terminal end is difficult to determine. However, we assume that the cysteine proteinase domain is 250 residues long and ends at residue No. 330 on the basis of the following criteria: (i) the molecular size of cysteine proteinase (220–250 residues long), (ii) the presence of a tryptophan cluster in the C-terminal region, and (iii) the distance between the active site histidine and the C-terminus (ca. 60 residues) [12, 21].

2.2.2 Ca^{2+} Binding Domain (Domain IV, Residues Nos. 561–705)

This domain is homologous to calmodulin and binds Ca^{2+}. Since CANP absolutely requires Ca^{2+} for activity, there must be some Ca^{2+} binding sites. The most typical Ca^{2+} binding site is the EF hand structure found in a variety of Ca^{2+} binding proteins like calmodulin and troponin C. The EF hand structure is a helix-loop-helix structure and Ca^{2+} binds to the loop region [25]. The search for the EF hand with its consensus sequence in the amino acid sequence of the 80 K subunit, reveals four candidates as shown in Figs. 1 and 2. These candidates have the following characteristics: (i) they exist consecutively with distinct spacings in the C-terminal region of the 80 K subunit, (ii) their sequences are highly homologous to the calmodulin sequence, and (iii) the secondary structures predicted from the amino acid sequence are helix-loop-helix. These features are typical of the EF hand structures found in calmodulin and many other Ca^{2+} binding proteins. Thus, these four candidates can be regarded as EF hand structures and the C-terminal region of about 150 residues in CANP is a calmodulin-like Ca^{2+} binding domain [12, 16, 21].

2.2.3 Other Domains

Two clear domains with completely different functions exist in the 80 K subunit, a cysteine protease domain (domain II) and a calmodulin-like Ca^{2+} binding domain (domain IV). There are at least two more domains, domain III which connects domains II and IV and domain I which precedes domain II. The functions of domains I and III are not clear from the sequence homology. They might be important for the interaction with the 30 K subunit or with an endogenous inhibitor or activator.

2.3 The Structure of the 30 K Subunit

The 30 K subunit of rabbit μ and mCANPs have been concluded to be identical from the results of the peptide maps on high performance liquid chromatography and the partial amino acid sequence analysis of their main peptides [26]. This fact is very im-

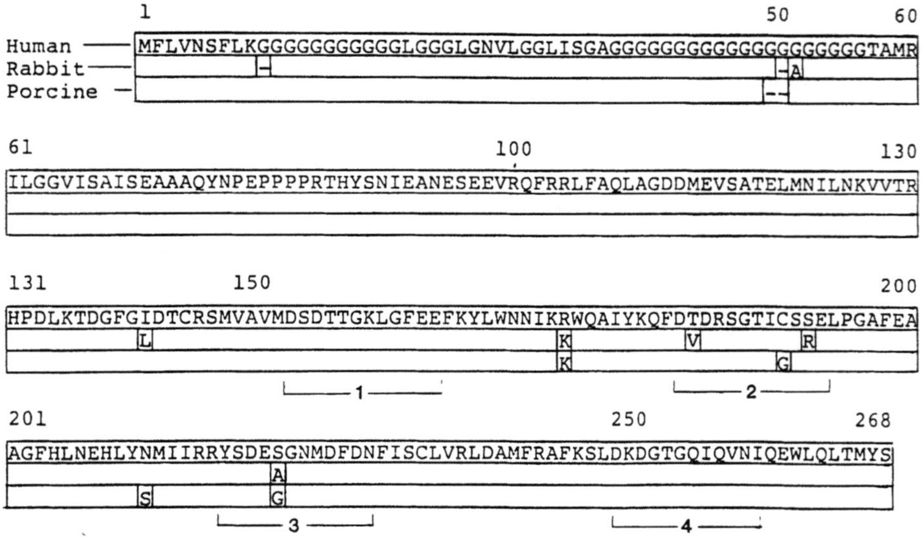

Fig. 4. Amino acid sequence of the 30 K subunit of CANP. For the sequences of the rabbit and porcine subunits, only the residues different from the human sequence are shown. Residue numbers are for human CANP. 1–4: loop region of EF hand structure

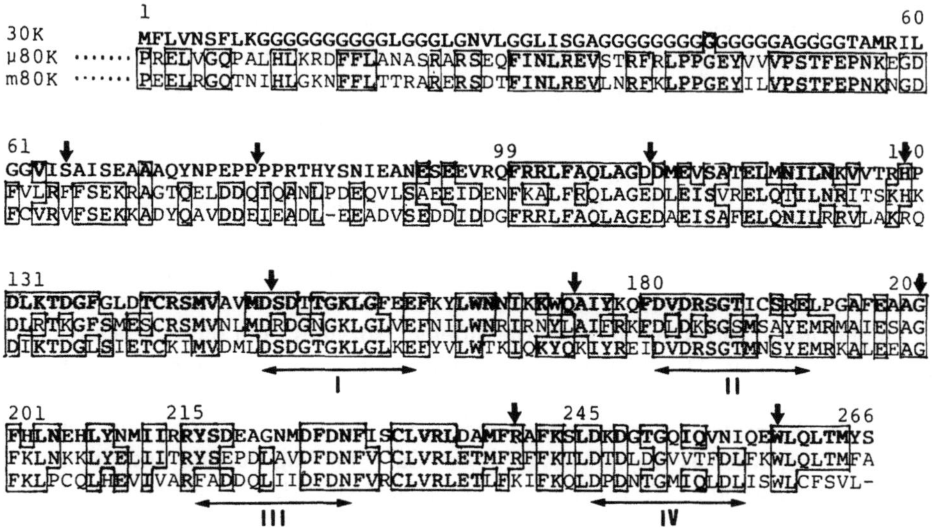

Fig. 5. Comparison of the amino acid sequence of the rabbit 30 K subunit with those of the C-terminal regions of the 80 K subunit of rabbit μ and mCANPs. Residue numbers are for the 30 K subunit. The intron points of the human 30 K subunit are shown by downward arrows. I–IV are the loop regions of EF hand structures

portant when the function of the 30 K subunit and the difference in Ca^{2+} sensitivity between μ and mCANPs are considered. The experiments were performed with the rabbit enzyme but the results are also applicable to CANPs from other sources.

The structures of the 30 K subunits of rabbit [18], human [22] and porcine [23] CANP have been determined mainly from the cDNA base sequence. They are composed of 266, 268 and 266 amino acid residues, respectively, and their molecular weights are about 28,000. As shown in Fig. 4, the structures of all 30 K subunits known at present are almost identical with sequence homologies of about 97% (see Table 1). The sequence of the 30 K subunit shows some novel structural features. First, the N-terminal 60–70 residues are extremely rich in glycine; the glycine content of this region is about 70%. Blocks of 10, 14 and 20 glycine residues are found. This N-terminal region is also rich in hydrophobic residues. Second, the C-terminal region of about 160 residues contains four consecutive EF hand structures and is regarded as a calmodulin-like Ca^{2+} binding domain. In addition, the sequence of this region (Nos. 99–266 of the rabbit 30 K subunit) is highly homologous to the corresponding C-terminal sequences of the μ and m 80 K subunit as shown in Fig. 5. This sequence homology is found only in the C-terminal region; the N-terminal sequence of the 30 K subunit is totally different from that of the 80 K subunit, though μ and m 80 K subunits show significant sequence homology in the upstream region [18, 20]. The C-terminal Ca^{2+} binding domains in the 80 K and 30 K subunits have evolved from the same ancestral gene as will be discussed later. Third, the existence of two regions rich in proline and glutamic acid or hydrophilic amino acids are noticeable.

As described above, the 30 K subunit is composed of at least two domains, the N-terminal glycine-rich hydrophobic domain and the C-terminal calmodulin-like

Ca^{2+} binding domain. The presence of the Ca^{2+} binding domain in the 30 K subunit indicates it is a regulatory subunit that controls the protease activity of the 80 K subunit [18, 20, 23]. As discussed later, the glycine-rich hydrophobic domain is important for the interaction with the cell membrane or other cellular organelles.

2.4 Comparison of the Structures of CANP

The amino acid sequences of various CANPs are compared and typical results are summarized in Table 1. The 80 K subunits of the same isozyme from different sources are highly homologous, but those of the different isozymes from the same source are similar but distinct. The chicken 80 K subunit is intermediate between the μ and m types. When the sequence homology for each domain is compared between μ and mCANPs, the protease domain shows the highest homology (70–80%) for various combinations, whereas the homology in the Ca^{2+} binding domain and domain I is the lowest (ca. 50%). In other words, μ and mCANPs have essentially the same protease domain and may differ only in Ca^{2+} sensitivity, which will be ascribed to differences in domains I and IV. The calmodulin-like Ca^{2+} binding domains in the C-terminal regions of the 80 K and 30 K subunits show about 50% identity in sequence.

Table 1. Sequence homology of CANP subunits

Combination	Homology (%)
80 K Subunit	
Human μ vs. Rabbit μ	95
Human m vs. Rabbit m	94
Human m vs. Human μ	61
Rabbit m vs. Rabbit μ	51
Chicken vs. Human μ	69
Chicken vs. Human m	66
30 K Subunit	
Human vs. Rabbit	97
Rabbit vs. Porcine	97
Porcine vs. Human	96
"Calmodulin" in 30 K and 80 K Subunits	
Rabbit 30 K vs. Rabbit μ 80 K	51
Rabbit 30 K vs. Rabbit m 80 K	50

3 Regulation of Protease Activity

3.1 Domain Structure and Regulation of Activity

Sequence homology of the protease domain to cysteine proteinases shows that the protease activity of CANP can be ascribed to this domain and that this domain should be active without Ca^{2+} when it is isolated from the whole CANP molecule [16, 20].

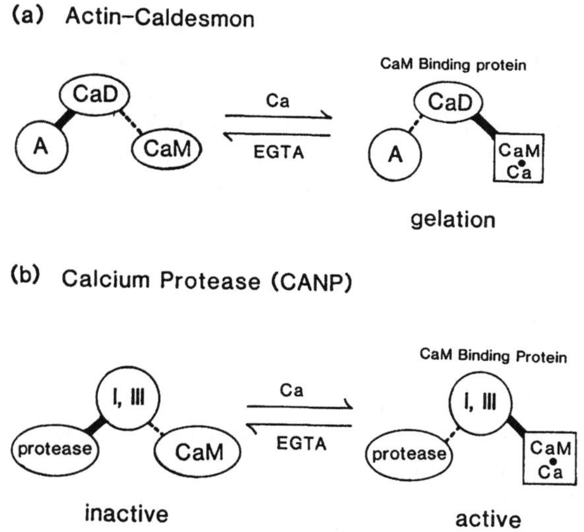

Fig. 6a and b. Flip-flop mechanism for the regulation of biological systems by calmodulin. Thick and dotted lines indicate strong and weak or no interaction between components. **a)** CaM, CaD, A: calmodulin, caldesmon and actin, respectively. **b)** Protease, I, III and CaM indicate protease domain (II), domains I and III, and calmodulin-like Ca^{2+} binding domain (IV), respectively

However, since CANP is inactive in the absence of Ca^{2+}, the activity of the protease domain is repressed or inhibited through interaction with other domains, e.g. domains I and III, in the whole CANP molecule. The function of two Ca^{2+} binding domains in the C-terminal regions of the 80 K and 30 K subunits is to repress the inhibition or stimulate the protease activity through the conformational change induced by binding of Ca^{2+}. The mechanism by which these two "calmodulins" function in the enzyme-activity regulation is not clear at present. The critical point is whether or not the protease domain is really active in the absence of Ca^{2+}. The answer is yes. We have recently isolated a "protease" from CANP which is active without Ca^{2+}.

There are at least eight potential Ca^{2+} binding sites in CANP, four each in the 80 K and 30 K subunits. Preliminary equilibrium dialysis experiments with chicken CANP indicate that six moles of Ca^{2+} bind to CANP with an apparent average dissociation constant of 75 μM [21]. Hence most of the Ca^{2+} binding sites predicted from the sequence actually bind Ca^{2+}.

The regulation of the protease activity of CANP by Ca^{2+} and calmodulin can be explained in terms of a flip-flop model proposed by Kakiuchi et al. [27]. In a model case shown in Fig. 6a, actin interacts with caldesmon in the absence of Ca^{2+}, and calmodulin has no effect. In the presence of Ca^{2+}, caldesmon binds to Ca^{2+}-calmodulin complex as a calmodulin binding protein and actin becomes free to initiate the gelation. In the case of CANP (Fig. 6b), the protease domain interacts with domain I and/or III in the absence of Ca^{2+} and the protease activity is repressed. With Ca^{2+}, domains I and/or III interact with "calmodulin"-Ca^{2+} as a calmodulin binding protein and the protease domain becomes free and active. Significant differences, however, exist between these two systems. In the former case, calmodulin and the calmodulin binding protein are both independent molecules separated from the target protein. In the latter, "calmodulin" and calmodulin binding proteins are covalently bound to the target protein, in this case protease domain.

Calmodulin usually exists in cytosol and activates the target enzyme by forming a complex upon binding Ca^{2+} [28]. In some cases, calmodulin is found as a subunit of the target enzyme. "Calmodulin" in the 30 K subunit of CANP is a subunit type. One "calmodulin" in the 80 K subunit is convalently linked to the target enzyme, a situation that has not been found before. The regulation of protease activity by covalently bound calmodulin means the absolute dependence of the CANP activity on Ca^{2+}. Considering that intracellular protease activity is potentially hazardous and often has lethal effect on cells, this strict regulation of CANP activity is quite reasonable.

3.2 Changes in Ca^{2+} Sensitivity

Judging from its Ca^{2+} sensitivity, CANP, especially mCANP, is not active under physiological Ca^{2+} concentrations, unless its Ca^{2+} sensitivity is increased. Since mCANP is the main molecular species in most cells, it is natural to assume that its Ca^{2+} sensitivity is raised by a certain mechanism to allow activity at physiological Ca^{2+} concentration. The Ca^{2+} sensitivity of CANP is modified by two methods, by metal ions like Mn^{2+} or Ba^{2+} [11] and by autolysis [11].

Activation of mCANP by mM Mn^{2+} or Ba^{2+} does not occur, but their presence increases the Ca^{2+} sensitivity of CANP so that it becomes active in the presence of µM Ca^{2+}. This activation by Mn^{2+} or Ba^{2+} will not occur in vivo because the required concentration of Mn^{2+} or Ba^{2+} can not be reached under physiological conditions. Mn^{2+} mimics the effect of Ca^{2+} in exposing the active site cysteine residue to the surface by inducing a conformation change in CANP [11]. It is not clear whether this type of activation of CANP is an artifact observed in vitro or whether some other effectors like Mn^{2+} exist in vivo.

Very rapid autolysis of mCANP occurs upon incubation in mM Ca^{2+}. Changes in the molecular weight and the activity of chicken CANP are summarized in Fig. 7 [11, 29-33]. The 80 K subunit changes into a 78 K species within 1 min even at 0 °C; 58 K species appears on longer incubation. The 30 K species also changes rapidly into a

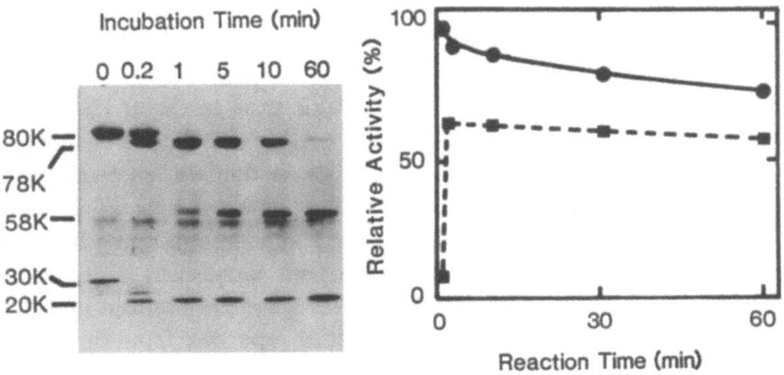

Fig. 7. Changes in the electrophoretic pattern and the Ca^{2+} sensitivity of chicken CANP during autolysis. Autolysis was performed at 0 °C in 2 mM Ca^{2+}. The total CANP activity and µCANP activity were measured at 2 mM and 100 µM Ca^{2+}, respectively

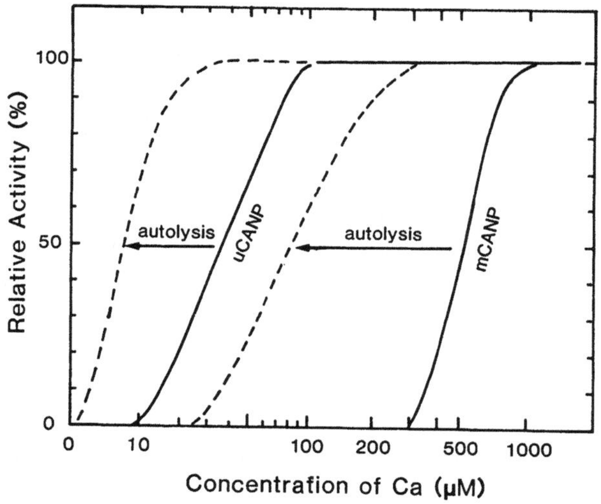

Fig. 8. Change in the Ca^{2+} sensitivity of rabbit μ and mCANPs by autolysis. Solid and broken lines denote the Ca^{2+} sensitivity before and after autolysis, respectively

20 K species that is stable on longer incubation [11,31,32]. The total activity of CANP gradually decreases during this autolysis. The μCANP activity at 100 μM Ca^{2+} is almost zero at the beginning but increases rapidly during autolysis and then declines in parallel with total CANP activity. Similar autocatalytic changes in the activity and electrophoretic patterns are observed with μ and mCANPs from various sources. The Ca^{2+} sensitivity of CANP increases significantly during autolysis. As shown in Fig. 8, the Ka values (Ca^{2+} concentration for 50% activity) for μ and mCANPs change from 30 μM to 7 μM and from 0.5 mM to 70 μM, respectively, on autolysis [29].

The analysis of autolyzed CANP indicates that only the N-terminal regions of the 80 K and 30 K subunits are modified while the C-terminal regions remain intact during autolysis. In the case of the chicken and rabbit μ and m 80 K subunits, 16, 27 and 19 amino acid residues are removed from the N-terminus, respectively [29], whereas 89 residues are removed from the N-terminal region of the rabbit 30 K subunit [29]. Thus, the glycine-rich hydrophobic domain in the 30 K subunit is completely removed by autolysis. As autolyzed CANP is still a dimer, the N-terminal domain of the 30 K subunit is not important for the association with the 80 K subunit [29,33]. A hybridization experiment of native and autolyzed CANP was performed to determine which subunit is responsible for the change in the Ca^{2+} sensitivity [29]. The subunits of native and autolyzed rabbit mCANPs were separated by gel filtration in 1 M NaSCN, and then two hybrid CANPs were reconstituted by removal of NaSCN following subunit exchange. One hybrid composed of the native 80 K subunit and the autolyzed 30 K subunit had the same Ca^{2+} sensitivity as that of native mCANP, whereas the other hybrid comprising the autolyzed 80 K subunit and the native 30 K subunit showed enhanced Ca^{2+} sensitivity and was active at μM Ca^{2+}. It is concluded that modification of the 80 K subunit results in the change of Ca^{2+} sensitivity. This conclusion is quite reasonable because the high degree of homology between the 30 K

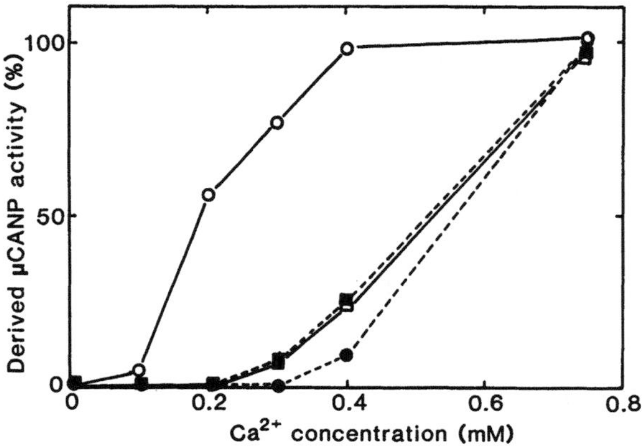

Fig. 9. Effect of phospholipid on the Ca^{2+} requirement for the autolysis of chicken CANP. Solid and broken lines are for native CANP and hybrid CANP devoid of the N-terminal region of the 30 K subunit, respectively. Open and filled symbols are with and without phospholipid

subunits of μ and mCANPs suggests that the difference in Ca^{2+} sensitivity of μ and mCANPs is primarily determined by their 80 K subunits. The 30 K subunit has apparently no effect on the activity and Ca^{2+} sensitivity of CANP.

Autolysis of mCANP can not occur in vivo judging from its Ca^{2+} sensitivity as measured in vitro. However, the concentration of Ca^{2+} required for autolysis is decreased to about one-tenth in the presence of phospholipid, e.g. phosphatidylinositol, as shown in Fig. 9 [30, 33]. The effect of phospholipid is observed with both μ and mCANPs, but not with CANP devoid of the N-terminal glycine-rich hydrophobic domain in the 30 K subunit [33]. The N-terminal glycine-rich hydrophobic domain is thus essential for the interaction with phospholipid, CANP binds to liposomes or reticulocyte membranes only in the presence of a μM level of Ca^{2+} [29, 33–35]. CANP interacts with the cell membrane through the glycine-rich hydrophobic domain in the 30 K subunit and autocatalytic activation of CANP occurs at the cell membrane [29, 35].

3.3 Activation Mechanism of CANP

As described above, autolyzed μ and mCANPs are active at physiological Ca^{2+} concentrations and this autocatalytic activation of CANP occurs at the cell membrane in the presence of phospholipid and Ca^{2+}. The activation mechanism is summarized in Fig. 10 [29].

CANP usually exists in the cytosol. When the cytosolic Ca^{2+} concentration is raised, CANP binds to the inner side of thc cell membrane through the N-terminal glycine-rich hydrophobic domain in the 30 K subunit (step 1). Autocatalytic activation, i.e. modification of the 80 K subunit, occurs at the cell membrane in the presence of phospholipid and a physiological concentration of Ca^{2+} (step 2). A proline-rich region in the 30 K subunit may be a hinge connecting the glycine-rich hydrophobic domain and the Ca^{2+} binding domain (IV′) essential for the association with the catalytic subunit [18]. A glutamic-acid rich hydrophilic region may terminate the hydro-

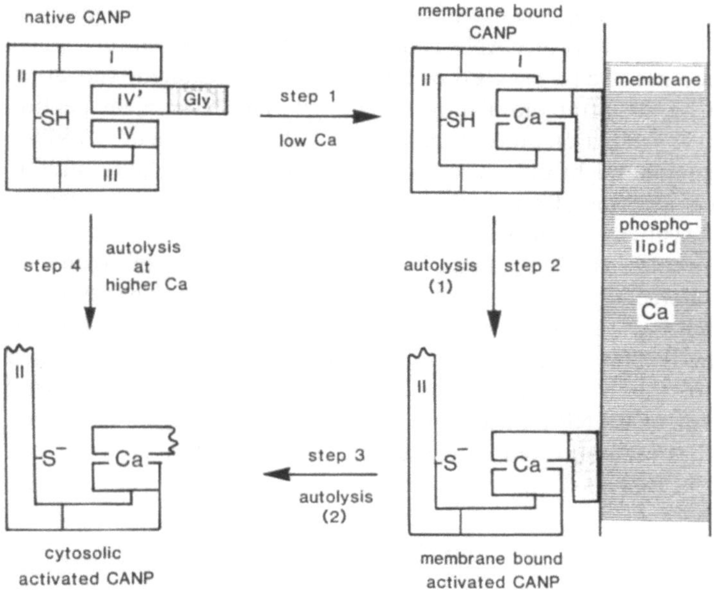

Fig. 10. Activation mechanism of CANP by autolysis. I–IV: domains in the 80 K subunit. IV′, Gly: calmodulin-like Ca^{2+} binding domain homologous to IV and glycine-rich hydrophobic domain in the 30 K subunit, respectively. SH, S^-: inactive and activated SH group of the active site cysteine residue, respectively

phobic interaction between the glycine-rich hydrophobic domain and the cell membrane [18]. CANP thus activated will function at the cell membrane. A part of the activated CANP is liberated from the membrane upon hydrolysis of the glycine-rich hydrophobic domain and becomes active in the cytosol (step 3). The N-terminal region of the 30 K subunit is always removed on autolysis in vitro together with the modification of the 80 K subunit. However, since this region interacts with the cell membrane, it may be resistant to autolysis in vivo.

Autolysis of CANP also occurs in the cytosol without phospholipid but it requires a much higher Ca^{2+} concentration (step 4). Thus the autolysis of μCANP, but not mCANP, occurs in the cytosol. Autolyzed μCANP should be active both in the cytosol. Autolyzed μCANP should be active both in the cytosol and at the cell membrane, but the activated mCANP will be active only at the cell membrane, because its Ca^{2+} sensitivity is not high enough to be active in the cytosol.

Although CANP is regarded as a cytosolic enzyme, the interaction of CANP with the cell membrane seems crucial for its activation and expression of activity.

In the above discussion, we assume that autolysis of CANP is required for the increase in Ca^{2+} sensitivity but the activity of native or unautolyzed CANP is not defined clearly. Pontremoli et al. proposed that dimeric CANP is an inactive proenzyme and that the 80 K subunit is activated by dissociation from the 30 K subunit and by autolysis [36]. The active "monomeric" enzyme is apparently different from the "dimeric" form discussed here. Inomata et al. [32] and Coolican et al. [31] analyzed the autolysis of CANP in the presence of a substrate and concluded that native

CANP appears to be a proenzyme because autolysis of CANP precedes the hydrolysis of the substrate. As the autolysis of CANP is independent of its concentration and occurs even in the presence of a large excess of substrate, the autocatalytic activation of CANP seems to be an intramolecular reaction [11,32]. The N-terminal peptide removed from the 80 K subunit on autolysis might block the active site of CANP or the removal of the peptide might induce an active conformation. The autocatalytic activation may be a conversion of a proenzyme into an active CANP rather than a change in the Ca^{2+} sensitivity of the active enzyme.

Ca^{2+} plays various roles in the activation of CANP. First, Ca^{2+} is involved directly or indirectly in the interaction between CANP and the cell membrane or phospholipid [30,33-35]. Second, Ca^{2+} exposes the buried active site cysteine to the surface of the enzyme molecule [11]. Third, the reactivity of the active site cysteine is increased greatly by addition of Ca^{2+} [11]. These effects of Ca^{2+} are generally ascribed to a conformational change of CANP induced upon Ca^{2+} binding, though a clear conformational change has been observed only in the second case.

3.4 *Inhibitor and Activator of CANP*

The activity of CANP is absolutely dependent on the presence of Ca^{2+}. Two protein factors that regulate the CANP activity in the presence of Ca^{2+} have been found; endogenous inhibitor and activator. CANP inhibitor, also known as calpastatin, was first identified in 1978 in extracts of bovine heart and rat liver [2,7]. The inhibitor is specific for CANP and requires Ca^{2+} for inhibition to occur. Subsequently, a similar inhibitor specific for CANP has been isolated from various sources [37-43]. The inhibitor preparations have several properties in common, e.g. heat stability, low content of aromatic amino acids, etc., but their molecular weights are significantly different. For example, the molecular weights of the inhibitor purified from human liver [37] and erythrocytes [43] are 107 K and 70 K, respectively, by SDS gel electrophoresis. Values of 24 K [38] to 172 K [42] are reported for the "monomer" form of the inhibitor from various sources. Further, multiple bands of the inhibitor are detected in extracts of mammalian organs by immunoblotting [43]. The molecular diversity of the inhibitor may arise from dissociation-association of the monomeric form of the inhibitor. The inhibitor often degrades into active smaller fragments during preparation or upon storage without significant change in the inhibitory activity [37]. It is not clear at present whether the various molecular species of the monomer originate from the fragmentation of a single large inhibitor [37,43]. The smallest inhibitor thus far identified has a molecular weight of about 20,000–30,000 and inhibits one mole of CANP [38]. Larger inhibitors presumably consist of repeats of this minimum inhibitory unit. However, different values have been reported for the stoichiometry of inhibition of CANP. Thus the 107 K inhibitor of human liver inhibits 5 moles of CANP [37], whereas the 107 K inhibitor of pig heart and 68 K inhibitor of pig erythrocytes [43] inhibit 8 and 5 moles of CANP, respectively. Formation of an inhibitor-CANP complex inhibits m and μCANPs to the same extent. The presence of the Ca^{2+} concentration required for CANP activity is also essential for the formation of the inhibitor-CANP complex [37-43]. Complexes with mCANP are formed at mM Ca^{2+} while μCANP is inhibited by the inhibitor at μM Ca^{2+}.

The inhibition is reversible. CANP and the inhibitor can be recovered in their native states by dissociation of the CANP-inhibitor complex under appropriate conditions [44]. The CANP inhibitor inhibits the association of CANP to the cell membrane by forming a complex [45]. This means that the inhibitor inhibits the CANP activity in two ways: by reducing the number of active CANP molecules and also by inhibiting the activation of additional CANP at the cell membrane.

Recently, additional CANP inhibitors have been found in human plasma, α-thiol proteinase inhibitors (α_1- and α_2-TPIs) [46]. These inhibitors inhibit not only CANP but cysteine proteinases in general. The cDNA cloning of α_2-TPI reveals that it corresponds to low molecular weight kininogen missing the kinin moiety [47]. It has been reported that the heavy chain of low and high molecular weight kininogens inhibits CANP [48, 49]. CANP exists in the cells and kininogen exists in plasma. Thus they do not interact under normal conditions, but the inhibition of CANP by kininogen may be important under certain pathological conditions such as inflammation [48].

DeMartino and Blumenthal first described a heat-stable protein factor in bovine brain cytosol which markedly activates both μ and mCANPs [6, 10, 11]. Recently, Takeyama et al. have partially purified a similar activator of CANP associated with bovine brain microsomal-insoluble elements [50]. Although the apparent cellular localization of these two factors is different, they are presumed to be identical, because their properties are virtually the same. The activator activates both μ and mCANPs about 10-fold without alteration of the affinity for Ca^{2+} or substrate. The molecular weight is about 15,000 and no loss of activity is observed after heating at 95 °C for 10 min. Although the nature and function of this activator remain unknown, it is worth noting that CANP cleaves cystoskeletal proteins preferentially and the activator is associated with such structural elements.

4 Gene Structure and Molecular Evolution of CANP

4.1 Gene Structure

The gene structures of the 80 K and 30 K subunits were analyzed to clarify the domain structure and molecular evolution of CANP. The gene for the chicken 80 K subunit is unique and about 10 kb long. It is composed of 21 exons of various sizes from 12 kb to 1.2 kb as shown in Fig. 11 [51]. Interesting features were revealed when this gene structure and the domain structure were considered. First, in the protease domain, the region around the active site Cys-108, which is highly conserved among cysteine proteinases, is split by the longest intron. Furthermore, three introns exist between the active site cysteine and histidine regions. The intron-exon structure does not correspond well to the functional unit in this domain. Since the gene structure of other cysteine proteinases has not been elucidated, it is not clear whether the features observed with the chicken 80 K subunit are common to other cysteine proteinases or unique to CANP. Second, in the calmodulin-like Ca^{2+} binding domain, each of the four EF hand structures is encoded by one exon. Thus the gene structure of this domain

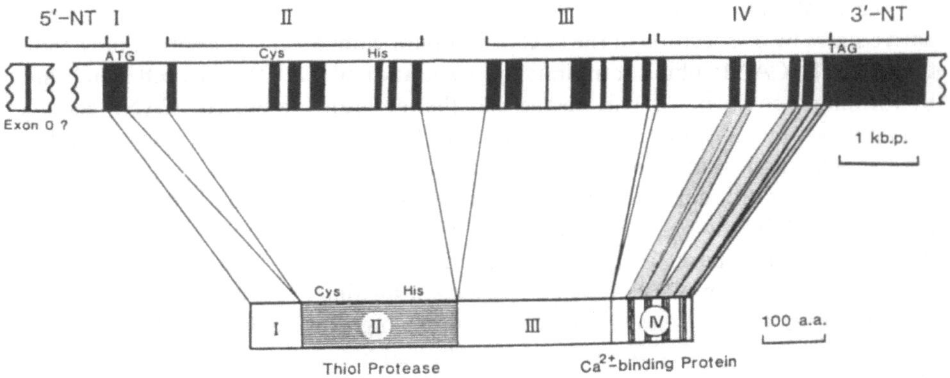

Fig. 11. Gene structure of the chicken 80 K subunit. Black and white regions are exons and introns, respectively. ATG, TAG: initiation and termination codon of the gene, respectively

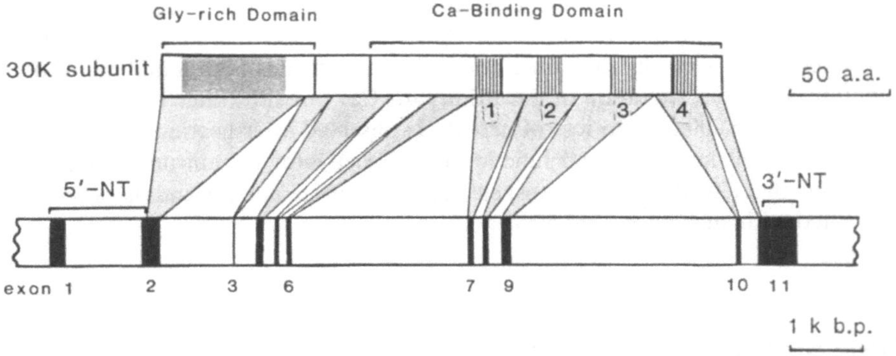

Fig. 12. Gene structure of the human 30 K subunit. Black and white regions are exons and introns, respectively. 1–4: EF hand structures. Lateral stripes: glycine rich region

shows good correlation with the functional unit. Moreover, intron break points are always found at or very close to the domain junctions. This confirms the domains defined on the basis of the amino acid sequence homology.

The gene structure of the human CANP 30 K subunit is summarized in Fig. 12 [52]. It is a unique gene about 10 kb long comprising 11 exons. The N-terminal glycine-rich hydrophobic domain is encoded by one exon, and in the C-terminal calmodulin-like domain, each of the four EF hand structures is encoded by one exon. This situation is exactly the same as that found in the corresponding domain in the 80 K subunit. The calmodulin-like domain in the 80 K and 30 K subunits are about 60% identical. The intron points in this region are exactly the same for the two genes in terms of their nucleotide and amino acid sequences (Fig. 5) [20, 52]. This clearly indicates that they have evolved from the same ancestral gene.

4.2 Molecular Evolution

It has been proposed from the sequence homology that the four consecutive EF hand structures in calmodulin have evolved from a primordial peptide with one Ca^{2+} binding helix-loop-helix structure by a two-step gene duplication (Fig. 13). CANP,

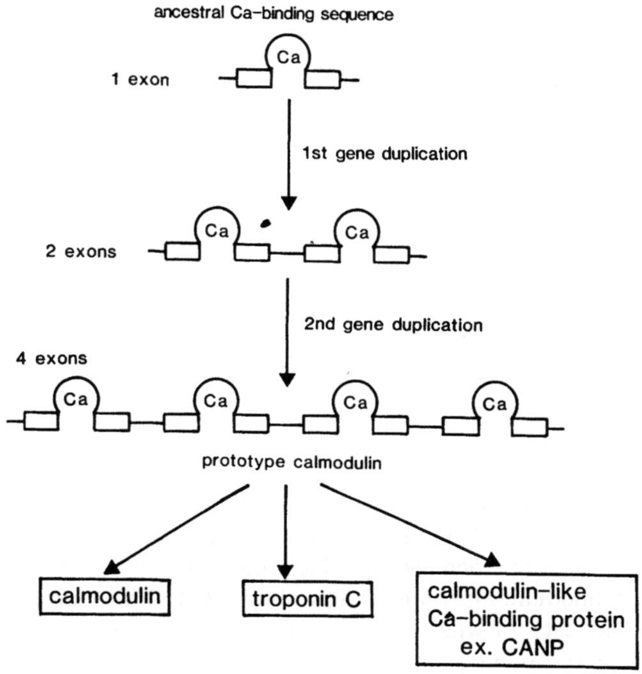

Fig. 13. Molecular evolution of the calmodulin family

like calmodulin, has four consecutive EF hand structures as potential Ca^{2+} binding sites in the C-terminal regions of both subunits. The genomic structure of CANP in these regions is consistent with the gene duplication theory and the structure of the prototype calmodulin gene is preserved. Thus, each of the four EF hand structures is encoded by one exon, and intron break points are generally located in the α-helical structures that border the Ca^{2+} binding loop.

In contrast, the gene structures of chicken [53] and rat [54] calmodulins are not consistent with this idea. Although the amino acid sequences are completely conserved in these two species, the positions of the introns are different, some of them being located in the middle of the Ca^{2+} binding loops. In the case of calmodulin, the original gene structure is lost probably due to gene rearrangement that occurred after the gene duplication.

The molecular evolution of CANP is summarized in Fig. 14. The 80 K subunit of CANP arose by fusion of four genes: the genes for calmodulin, cysteine proteinase,

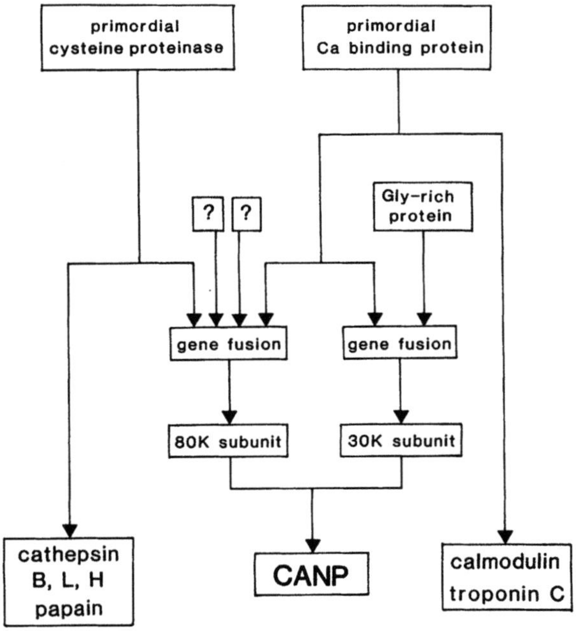

Fig. 14.
Molecular evolution of CANP

and two as yet unidentified genes. The 80 K subunit belongs to the superfamilies of calmodulins and cysteine proteinases, while in the 30 K subunit, the same calmodulin gene fused to a gene coding for a glycine-rich hydrophobic protein. The 30 K subunit is thus also a member of the calmodulin superfamily. Accordingly, CANP was formed by association of two subunits each of which arose from the fusion of genes with completely different evolutionary origins and functions. CANP is one of the most interesting proteins from the aspect of molecular evolution.

5 Biological Functions of CANP

5.1 Distribution of CANP

Comparative biochemical studies on CANP are scarce though CANP has been isolated from mammals, chicken, fish, octopus, cuttle fish, crab, crayfish, etc. [2,6,11]. CANP has not been found in plant, bacteria or yeast. Distribution of CANP along the phylogenetic tree must await further studies. CANP exists widely in various tissues and cells of higher animals as revealed by immunological methods and activity measurements [6]. Langerhans islets and red cells contain only μCANP [55], but in most other cells, both μ and mCANPs are detected. Essentially the same results have been obtained from determinations of the amount of mRNA for CANP [56]. Much higher amounts of mRNA have been detected in ovary, testis, and eye than in muscle and spleen [56]. For the most part, μ and mCANP exist in the cytosol [57], but the distribution of CANP is presumably affected by various factors, e.g. the cell cycle or the state of differentiation. Indeed, in rat myoblasts [58], immunofluorescence studies indicate that mCANP is predominantly associated with the plasma membranes of fusion competent cells, but in non-fusing cells, the distribution is generally cytosolic and diffuse. On the other hand, μCANP is localized in the cytoplasm of all the myoblasts. Thus mCANP may play a role in the Ca^{2+}-mediated fusion of myoblasts.

5.2 Natural Substrates for CANP and Their Biological Significance

CANP hydrolyzes a limited number of proteins in their native states to large fragments only and not to small peptides or amino acids [8,11,15]. Limited proteolysis of protein substrates is the most significant feature of the action of CANP. CANP shows little or no activity towards various small peptides that are readily cleaved by other proteinases. The substrate specificity of CANP seems to be determined by the conformation or length of a peptide chain rather than by some specific amino acid residues or sequence. However, the hydrolyzed peptide bonds usually have hydrophobic residues at their P_2 sites.

The various protein substrates hydrolyzed by CANP can be classified into four groups: (i) enzymes, especially kinases and phosphatases, (ii) muscle proteins, (iii) receptors for hormones or growth factors, and (iv) cytoskeletal proteins or membrane proteins.

Phosphorylase b kinase, protein kinase C, pyruvate kinase, phosphorylase phosphatase, fructose 1,6-bisphosphatase, polycation-stimulated protein phosphatase, phospholipase C, glycogen synthetase, and tryptophan hydroxylase are hydrolyzed by CANP without significant change in enzyme activity [6,10,11]. In most cases, the enzyme activity is increased upon hydrolysis. Protein kinase C and phospholipase C are both activated by CANP. Further, CANP binds specifically to the cell membrane in the presence of Ca^{2+} and is activated by phospholipid and Ca^{2+} at the cell membrane [33-35]. From these results, CANP is assumed to have some role in the signal transduction of cells. Pontremoli et al. reported the activation of CANP at the cell membrane of human neutrophils treated with phorbol ester (PMA) or formylmethionylleucylphenylalanine (fMLP). Activated CANP converts membrane bound protein kinase C to the soluble Ca^{2+}-phospholipid independent form [59]. The activation of CANP and protein kinase C, is essential for exocytosis of neutrophils promoted by PMA or fMLP [60].

Among muscle proteins, troponin, tropomyosin, myosin heavy chains and C-protein are hydrolyzed easily, but myosin light chains, M protein, actin and α-actinin are resistant to hydrolysis [4,10,11]. Ca^{2+} induces the degradation of myofibrillar proteins, a process that is specifically inhibited by E64 or its derivative, a specific potent inhibitor of CANP [4,10]. Further, CANP removes the Z-band very quickly upon incubation with myofibrils [4]. From these facts CANP is assumed to initiate the degradation of muscle proteins. It has also been reported that CANP may have some role in tissue resorption during uterin involution [61].

CANP degrades steroid hormone receptors for glucocorticoid, progesterone, and estrogen as well as growth factor receptors like epidermal growth factor (EGF) and platelet derived growth factor (PDGF) [62] to specific sizes without significant changes in apparent affinity to hormones or growth factors. PDGF and EGF receptors retain their kinase activity after hydrolysis. The biological implications of the hydrolysis of these receptors remain to be established.

Neurofilaments, filamin, fodrin, vimentin, desmin, tubulin, band 2.1, 3, 4.1 and MAPs are hydrolyzed by CANP [10,11]. Degradation of neurofilaments at the nerve ending is believed to be mediated by CANP [10]. CANP may be involved in various cellular events in which changes in membrane and cytoskeleton proteins play important roles, e.g. cell fusion and cell division. The number of glutamate receptors in rat hippocampus synaptic plasma membrane is increased after treatment with CANP. This is due to the hydrolysis of fodrin which regulates the state of the glutamate receptors [63]. CANP might have some correlation with memory.

The search for natural substrates of CANP has revealed its interesting and peculiar substrate specificity. Although these results were mainly obtained by in vitro experimentation, it is impossible to explain them collectively by assuming that CANP has a certain specific function. Presumably, CANP plays some common, fundamental role in all cells while also having special functions specific to various types of cells.

5.3 Correlation Between CANP and Disease

CANP is alway in contact with cytosolic protein substrates. If CANP were active at all times, these proteins would be constantly hydrolyzed resulting into serious cellular damage. Consequently, CANP must be in an inactive state and become active only when its activity is required by the cells. Active CANP must become inactive again immediately after the completion of the function. In other words, an activation-inactivation cycle for CANP exists in cells. Ca^{2+} is the most important factor in the regulation of this cycle, but activator, inhibitor or effector species that affect Ca^{2+} sensitivity also have a large effect on the modulation of CANP activity.

If this regulation system does not work properly, abnormal increase or decrease of CANP activity results, a situation that leads to various diseases. Muscular dystophy is probably attributable to an abnormal increase in CANP activity. In fact, a significant increase in CANP activity is observed in muscular dystrophy [1,4]. The increase in CANP activity does not necessarily mean a net increase in the amount of CANP. A decrease in the amount of inhibitor, changes in its localization or changes in the activation of CANP might result in an increase in CANP activity. If it were possible to reduce the enhanced CANP activity, the rate of degradation of muscle proteins in muscular dystrophy might be slowed down, although this might have no therapeutic value. Based on this notion, E64 and its derivatives [64], specific inhibitors of CANP and other cysteine proteinases, are now being tested as drugs for muscular dystrophy.

Wallerian degradation is another disease related to CANP [10]. It is presumably due to the degradation of neurofilaments by CANP activated by Ca^{2+} influxed from outside through a denatured membrane. The atrophy of the myocardium in myocardiac infarction may be attributable to CANP. Only a few diseases caused by the abnormal action of CANP are known. CANP, as a house keeping enzyme [52], may perform indispensable basic functions common to various cells. If so, then abnormal increases in CANP activity would be fatal to the organism and, therefore, CANP-related diseases may hardly exist.

6 Concluding Remarks

Various properties of CANP, i.e., protease activity, Ca^{2+} sensitivity and interaction with the membrane can be explained in terms of the domain structure of CANP. Nevertheless, the biological function is not yet clear. The novel structural features of CANP have had a strong impact on studies of the biological function of CANP. It is highly likely that CANP is involved in the control of various Ca^{2+}-mediated cellular processes. In particular, the events that take place at the cell membrane, i.e., stimulus-response coupling, has attracted special attention in conjunction with the role of Ca^{2+}. Although some circumstantial evidence has been obtained, none proves the direct participation of CANP in this process. It is hoped that the biological function of CANP will be clarified in the near future by the various new approaches that have been initiated by the results of the structural studies.

7 References

1. Imahori, K.: Characterization of calcium activated neutral protease (CANP), in: Muscular Dystrophy (ed.) Ebashi, S., p. 253, Tokyo, University of Tokyo Press 1980
2. Waxman, L.: Calcium-activated proteases in mammalian tissues, in: Methods in Enzymology (ed.) Lorand, L., Vol. *80*, p. 665 New York, Academic Press 1981
3. Murachi, T., Tanaka, K., Hatanaka, M., Murakami, T.: Adv. Enzyme Regul. *19*, 407 1981
4. Ishiura, S.: Life Sci. *29*, 1079 1981
5. Imahori, K.: Calcium-dependent neutral protease: its characterization and regulation, in: Calcium and Cell Function (ed.) Cheung, W. Y., Vol. *3*, p. 473, New York, Academic Press 1982
6. Murachi, T.: Intracellular Ca^{2+} protease and its inhibitor protein: calpain and calpastatin, in: Calcium and Cell Function (ed.) Cheung, W. Y., Vol. *4*, p. 377, New York, Academic Press 1982
7. Murachi, T.: Trends Biochem. Sci. *8*, 167 1983
8. Kay, J.: Ca^{2+}-activated proteases, protein degradation and muscular dystrophy, in: Proteases: Potential Role in Health and Disease (eds.) Heidland, A., Hore, H., p. 519, New York, Plenum 1984
9. Goll, E., Shannon, J. D., Edmunds, T., Sathe, S. K., Kleese, W. C., Nagainis, P. A.: Properties and regulation of the Ca^{2+}-dependent proteinase, in: Calcium Binding Proteins (ed.) de Bernard, B., Sottocassa, G. L., Sandri, G., Carafoli, E., Taylor, A. N., Vanaman, T. C., Williams, R. J. P., p. 19, Amsterdam, Elsevier Science 1983
10. Zimmerman, U.-J. P., Schlaepper, W. W.: Prog. Neurobiol. *23*, 63 1984
11. Suzuki, K., Kawashima, S., Imahori, K.: Structure and function of Ca^{2+}-activated protease, in: Calcium Regulation in Biological Systems (eds.) Ebashi, S., Endo, M., Imahori, K., Kakiuchi, S., Nishizuka, Y., p. 213, New York, Academic Press 1984

12. Suzuki, K., Ohno, S., Emori, Y., Imajoh, S., Kawasaki, H., Kisaragi, M.: Calcium-activated neutral protease (CANP) is a chimeric protein with distinct functional domains, in: The Biological Role of Proteinases and Their Inhibitors in Skin (eds.) Ogawa, H., Lazarus, G. S., Hopsu-Havu, V. K., p. 111, Tokyo, University of Tokyo Press 1985
13. Guroff, G.: J. Biol. Chem. *239*, 149 1964
14. Inomata, M., Nomoto, M., Hayashi, M., Nakamura, M., Imahori, K., Kawashima, S.: J. Biochem. *95*, 1661 1984
15. Sasaki, T., Kikuchi, T., Yumoto, N., Yoshimura, N., Murachi, T.: J. Biol. Chem. *259*, 12489 1984
16. Ohno, S., Emori, Y., Imajoh, S., Kawashima, H., Kisaragi, M., Suzuki, K.: Nature *312*, 566 1984
17. Emori, Y., Kawasaki, H., Sugihara, H., Imajoh, S., Kawashima, S., Suzuki, K.: J. Biol. Chem. *261*, 9465 1986
18. Emori, Y., Kawasaki, H., Imajoh, S., Kawashima, S., Suzuki, K.: J. Biol. Chem. *261*, 9472 1986
19. Aoki, K., Imajoh, S., Ohno, S., Emori, Y., Koike, Y., Kosaki, G., Suzuki, K.: FEBS Lett. *205*, 313 1986
20. Suzuki, K., Emori, Y., Ohno, S., Imajoh, S., Kawasaki, H., Miyake, S.: Biomed. Biophys. Acta *110*, 633 1986
21. Suzuki, K., Ohno, S., Emori, Y., Imajoh, S., Kawasaki, H.: J. Prot. Chem. *6*, 7 1987
22. Ohno, S., Emori, Y., Suzuki, K.: Nucleic Acid Res. *14*, 5559 1986
23. Sakihama, T., Kakidani, H., Zenita, K., Yumoto, N., Kikuchi, T., Sasaki, T., Kannagi, R., Nakanishi, S., Ohmori, M., Takio, K., Titani, K., Murachi, T.: Proc. Natl. Acad. Sci., USA *82*, 6075 1984
24. Takio, K., Towatari, T., Katunuma, N., Teller, D. C., Titani, K.: Proc. Natl. Acad. Sci., USA *80*, 3666 1983
25. Tufty, R. M., Kretsinger, R. H.: Science *187*, 167 1975
26. Kawasaki, H., Imajoh, S., Kawashima, S., Hayashi, H., Suzuki, K.: J. Biochem. *99*, 1525 1986
27. Kakiuchi, S., Sobue, K.: Trends Biochem. Sci. *8*, 59 1983
28. Klee, C. B., Vanaman, T. C.: Adv. Prot. Chem. *35*, 213 1982
29. Imajoh, S., Kawasaki, H., Suzuki, K.: J. Biochem. *100*, 633 1986
30. Coolican, S. A., Hathaway, D. R.: J. Biol. Chem. *259*, 11627 1984
31. Coolican, S. A., Haiech, J., Hathaway, D. R.: J. Biol. Chem. *261*, 4170 1986
32. Inomata, M., Hayashi, M., Nakamura, M., Imahori, K., Kawashima, S.: J. Biochem. *98*, 407 1985
33. Imajoh, S., Kawasaki, H., Suzuki, K.: J. Biochem. *99*, 1281 1986
34. Pontremoli, S., Salamino, F., Sparatore, B., Michetti, M., Sacco, O., Melloni, E.: Biochim. Biophys. Acta *831*, 335 1985
35. Melloni, E., Pontremoli, S., Michetti, M., Sacco, O., Sparatore, B., Salamino, F., Horecker, B. L.: Proc. Natl. Acad. Sci., USA *82*, 6435 1985
36. Pontremoli, S., Melloni, E., Sparatore, B., Salamino, F., Michetti, M., Sacco, O., Horecker, B. L.: Biochem. Biophys. Res. Commun. *128*, 331 1985
37. Imajoh, S., Kawasaki, H., Kisaragi, M., Mukai, M., Sugita, H., Suzuki, K.: Biomed. Res. *5*, 481 1984
38. Nakamura, S., Inomata, M., Hayashi, M., Imahori, K., Kawashima, S.: J. Biochem. *96*, 1399 1984
39. Nakamura, M., Inomata, M., Hayashi, M., Imahori, K., Kawashima, S.: J. Biochem. *98*, 757 1985
40. Melloni, E., Salamino, F., Sparatore, B., Michetti, M., Pontremoli, S., Horecker, B. L.: Arch. Biochem. Biophys. *232*, 513 1984
41. DeMartino, G. N., Croall, D. E.: Arch. Biochem. Biophys. *282*, 713 1984
42. Lepley, R. A., Pampusch, M., Dayton, W. R.: Biochem. Biophys. Acta *828*, 95 1985
43. Takano, E., Kitahara, A., Sasaki, T., Kannagi, R., Murachi, T.: Biochem. J. *235*, 97 1986
44. Imajoh, S., Suzuki, K.: FEBS Lett. *187*, 47 1985
45. Melloni, E., Salamino, F., Sparatore, B., Michetti, M., Pontremoli, S.: Biochem. Intern. *8*, 477 1984
46. Sasaki, M., Taniguchi, K., Suzuki, K., Imahori, K.: Biochem. Biophys. Res. Commun. *110*, 256 1983
47. Ohkubo, I., Kamachi, K., Takasawa, T., Shiokawa, H., Sasaki, M.: Biochemistry *23*, 5691 1984
48. Higashiyama, S., Ohkubo, I., Ishiguro, H., Kunimatsu, M., Sawaki, K., Sasaki, M.: Biochemistry *25*, 1669 1986

49. Salvesen, G., Parkes, C., Abrahamson, M., Grubb, A., Barrett, A. J.: Biochem. J. *234*, 429 1986
50. Takeyama, Y., Nakanishi, H., Uratsuji, Y., Kishimoto, A., Nishizuka, Y.: FEBS Lett. *194*, 110 1986
51. Emori, Y., Ohno, S., Tobita, M., Suzuki, K.: FEBS Lett. *194*, 249 1986
52. Miyake, M., Emori, Y., Suzuki, K.: Nucleic Acid Res. *14*, 8805 1986
53. Simmen, R. C., Tanaka, T., Tsui, K. F., Putkey, J. A., Scott, M. J., Lai, E. C., Means, A. R.: J. Biol. Chem. *260*, 907 1985
54. Nojima, H., Sokabe, H.: J. Mol. Biol. *193*, 439 1987
55. Kitahara, A., Ohtsuki, E., Kannagi, R., Murachi, T.: FEBS Lett. *184*, 120 1985
56. Suzuki, K., Ohno, S., Imajoh, S., Emori, Y., Kawasaki, H.: Biomed. Res. *6*, 323 1985
57. Yoshimura, N., Hatanaka, M., Kitahara, A., Kawaguchi, N., Murachi, T.: J. Biol. Chem. *259*, 9847 1984
58. Schollmeyer, J. E.: Exp. Cell Res. *163*, 413 1986
59. Melloni, E., Pontremoli, S., Michetti, M., Sparatore, B., Horecker, B. L.: J. Biol. Chem. *261*, 4101 1986
60. Pontremoli, S., Melloni, E., Michetti, M., Sacco, O., Salamino, F., Sparatore, B., Horecker, B. L.: J. Biol. Chem.: *261*, 8309 1986
61. Elce, J. S., Baenziger, J. E., Young, D. C. R.: Biochem. J. *220*, 507 1984
62. Ek, B., Heldin, C.-H.: Eur. J. Biochem. *155*, 409 1986
63. Siman, R., Baudry, M., Lynch, G.: Nature *313*, 225 1985
64. Hanada, K., Tamai, M., Adachi, T., Oguma, K., Kashiwagi, K., Ohmura, S., Kominami, E., Towatari, T., Katunuma, N.: Characterization of the three new analogs of E-64 and their therapeutic application, in: Protease Inhibitors. Medical and Biological Aspects (eds.) Katunuma, N., Umezawa, H., Holzer, H., p. 25, Tokyo/Heidelberg, Japan Sci. Soc. Press/-Springer 1983

Microbial Iron Transport
Structure and Function of Siderophores

Volkmar Braun and Günther Winkelmann

Mikrobiologie, Universität Tübingen,
Auf der Morgenstelle 28, D-7400 Tübingen/Germany

The extreme insolubility of iron limits the supply of most organisms by this metal which forms an essential part of enzymes active in the energy-generating electron transport chains and of the intermediary metabolism. The article emphasizes the intricate mechanisms by which microbes, bacteria as well as fungi, overcome iron deficiency imposed by the human host. In humans extracelluar iron is almost exclusively bound to transferrin and lactoferrin. To withdraw iron from these proteins most microbes secrete siderophores which exert an extremely high affinity and selectivity for iron (III). In addition, they posess unique transport systems by which the iron(III)-siderophores are actively transported into the microbial cells. Upon infection the human host actively reduces the iron available to microbes. The unique properties render iron (III) a suitable target for the host to control microbial growth ("nutritional immunity"). Interference with siderophore biosynthesis is considered as a therapeutic means to treat microbial infections. Highly effective antibiotics are known which contain a siderophore as carrier of the antimicrobially active moiety. A siderophore is used to remove iron which has accumulated in the body as a consequence of iron overload diseases.

Progress in Clinical Biochemistry
and Medicine, Vol. 5

1 Introduction

Iron is an essential element in almost all organisms. In bacteria it is mainly contained in the redox centers of enzymes located in the cytoplasmic membrane and in the cytoplasm. The iron content of *Escherichia coli* was found to be in the order of 10^6 ions per cell [1,2,3]. This means that a culture of 10^9 cells per ml requires 10^{15} Fe^{3+} ions per generation. However, after addition of iron salts to a culture medium at pH 7 under aerobic conditions. Fe^{3+} forms insoluble polynuclear ferric oxidohydroxides which cannot be taken up by microorganisms [4]. The solubility product constant is close to 10^{-36}, resulting in about 10^3 free iron ions per ml. The gap of 10^{12} between supply and demand is overcome by synthesis and release of Fe^{3+} ligands, designated siderophores. Siderophores are small organic molecules of several hundred Daltons in size, in contrast to the iron transport and storage compounds of higher organisms which are proteins. The affinity and specificity of the siderophores for Fe^{3+} is very high. Their formation constants are in the range of 10^{32} for the hydroxamate siderophores, and 10^{52} for the bacterial phenolate siderophore enterochelin (enterobactin) [5]. A physiological more meaningful conversion of the formation constants into pM values is discussed later (see Table 2). Microbial siderophores effectively withdraw Fe^{3+} present in equilibrium with the oxidohydroxide polymer, and with transferrin in the serum, or lactoferrin in secretory fluids.

The iron transport systems are unusual when compared to the transport of other nutrients. In contrast to amino acids, sugars, nucleosides and inorganic ions, iron is not transported as such but as a siderophore complex (under aerobic conditions). The Fe^{3+}-siderophores are taken up into the microorganisms via specific transport systems. Fe^{3+} is released by reduction to Fe^{2+}. The freed siderophores are either reused or destroyed. To cope with this unique condition microbial cells have invented new transport mechanisms.

Aside from interest in the various ways microbes overcome the extreme iron shortage owing to its insolubility, iron exerts strong regulatory effects on the synthesis of a number of bacterial exotoxins (for example on diphtheria, shiga, and tetanus toxins), some hemolysins and exoproteins of *Pseudomonas aeruginosa*. Iron supply systems and iron regulatory devices are factors which contribute to the virulence of a microbial pathogen [6,7]. Moreover, Desferal is a derivative of a microbial siderophore used to treat iron overload diseases [8]. Also, some root-colonizing microbes secrete siderophores which suppress growth of pathogenic microbes by iron withdrawal. Still another field of interest comes from similarities between Fe^{3+} and Pu^{4+}. Siderophores can serve as leads for synthetic chelating compounds of Pu^{4+} and other actinides [5].

In the following, the essential aspects of iron transport into *Escherichia coli* will be discussed. The most detailed knowledge has been gained mainly through the employment of molecular genetics. A brief review of the iron supply systems of other bacteria will follow. The final chapters deal with the large variety of siderophores formed by fungi and their iron transport systems.

2 Iron Transport Systems of Escherichia Coli

2.1 General Features

Study of iron transport into *E. coli* involves consideration of 4 compartments: the outer membrane, the periplasm, the cytoplasmic membrane and the cytoplasm. Solutes not larger than 600 Daltons pass across the outer membrane by diffusion through waterfilled protein channels. Since the molecular weight of most of the iron-siderophores exceeds 600 Daltons, *E. coli*, and related Gram-negative bacteria, have developed special mechanisms to transport iron across the outer membrane. The periplasm does not form a diffusion barrier. It contains so-called binding proteins, which are essential constituents of transport systems for certain sugars and amino acids. The cytoplasmic membrane is impermeable for all charged and most of the uncharged but hydrophilic solutes unless it contains specific transport proteins.

Five Fe^{3+} transport systems are known which can be expressed by *E. coli*. Only one siderophore, enterochelin or enterobactin, is regularly synthesized by the organism itself. Synthesis of a second siderophore, aerobactin, is encoded on large plasmids, mostly ColV plasmids, although the genes for aerobactin biosynthesis and transport have also been found on the chromosome. In case of plasmid coding, only those strains are able to acquire Fe^{3+}-aerobactin which bear the plasmid. Curiously, for 2 siderophores which are only produced by fungi, ferrichrome and coprogen, *E. coli* expresses 2 transport systems. Fe^{3+}-dicitrate also supplies iron. The transport system is induced by iron and citrate present in the growth medium. Citrate is not secreted in amounts (0.1 mM) sufficient for induction. It must be supplied by the ambient in which *E. coli* grows.

The existence of 5 different Fe^{3+} transport systems may reflect the different iron sources to which *E. coli* is exposed in various environments. There is evidence for two additional Fe^{3+} transport systems (see later). Furthermore, under anaerobic conditions, *E. coli* transports Fe^{2+}. The genetic locus responsible for Fe^{2+} transport has been named *feo* and was mapped close to *pncA* at 39 min on the *E. coli* linkage map (K. Hantke, unpublished).

The Fe^{3+} transport systems have been defined genetically and biochemically. Mutants were isolated which were deficient in one of the iron transport systems. Other mutants were impaired in all Fe^{3+} transport systems. Characterization of these mutants enabled assigning genetic loci to single transport systems and to components common to all Fe^{3+} transport systems. Outer membrane proteins involved in Fe^{3+} transport could be identified by staining after polyacrylamide gel electrophoretic separation. They were related to iron transport by virtue of their overexpression under iron-limiting growth conditions. Their absence in certain transport-deficient mutants assigned them to one of the 5 transport systems. Additional iron transport proteins were not expressed in sufficient amounts at low iron conditions but required amplification of their structural genes to be identified.

The five iron transport systems were identified in the following way. Enterochelin was secreted during growth in low iron media and turned the culture pinkish-red upon addition of iron. The structure of enterochelin, its biosynthesis and transport is largely elucidated [9].

The transport system via enterochelin was inactivated by isolating mutants devoid of enterochelin synthesis. Growth under iron-limitation then depended on the addition of siderophores, for example ferrichrome, coprogen, or citrate. Strains able to express the Fe^{3+}-aerobactin transport system were also characterized in an enterochelin-negative background. A convenient way to create iron-limitation was the addition of 0.2 mM 2,2′ dipyridyl whose iron complex does not serve as an iron source.

The most efficient method of determining structural genes, the proteins they encode and their functions, is cloning of DNA fragments from gene loci, previously defined by mutation, on multicopy plasmids. Complementation of independently isolated mutants by subclones containing segments of the total DNA fragment permits physical mapping of the mutation sites. Expression of the cloned DNA fragments behind strong promoters in maxi- or minicells, or in an in vitro transcription/translation system, correlates DNA fragments with polypeptides and phenotypes. In the case of the iron transport systems of *E. coli*, radioactively labeled proteins were correlated with the biosynthesis of siderophores and the transport of Fe^{3+}-siderophores which were formed in amounts too low to be identified in normal cells by staining of polyacrylamide gels. In addition, the proteins could be localized in the cells after subfractionation into cellular compartments.

2.2 Ferrichrome Transport

Fe^{3+} as ferrichrome complex is taken up by *E. coli* with an apparent K_M of 0.06 μM and a V_{Max} of 50 pmol/mg cell dry weight [10]. Dissection of the transport system into its constituent parts was greatly facilitated by the antibiotic albomycin, which is transported into cells via the same route as ferrichrome. In albomycin, Fe^{3+} is bound by three residues of N^5-acetyl-N^5-hydroxyornithine as in ferrichrome (Fig. 1). It has an open structure since it lacks the three glycine residues which together with the ornithine derivatives form the closed hexapeptide ring in ferrichrome. Instead, albomycin contains a L-serine which forms the link to 6-amino-6-deoxy-4-thio-L-glycero-α-L-ido-heptofuranuronic acid which is substituted in the 1-position by a pyrimidine derivative [11,12]. The nuceloside is presumably the antibiotically active portion of albomycin which is carried into the cells by the Fe^{3+} ligand. In order to be active, the Fe^{3+} ligand has to be cleaved off by peptidase N [13]. In mutants lacking peptidase N activity, albomycin serves as a siderophore. The other albomycin-resistant mutants are defective in transport. Four contiguous transport genes have been mapped and sequenced [14,15,16,17]. They were named *fhuA* (former designation *tonA*), *fhuC*, *fhuD* and *fhuB* according to their order on the *E. coli* chromosome at 3 min.

The FhuA protein, which forms the receptor protein in the outer membrane, is encoded by 2241 nucleotides [15] and is synthesized as a precursor forms with 33 amino acids at the aminoterminus which are cleaved off during export from the cytoplasm to the outer membrane. The calculated molecular weight of the mature form is 78,992. The protein does not contain extended regions made up to hydrophobic amino acids. Nevertheless, the FhuA protein is firmly embedded in the outer membrane. The FhuA protein could only be isolated in quantities which were of an electrophoretically homogeneous form from cells containing three *fhuA* gene copies on a thermo-amplifiable multicopy plasmid [18]. Chromosomally encoded FhuA protein was only purified in low amounts [19]. Upon strong overexpression of the FhuA protein, synthesis

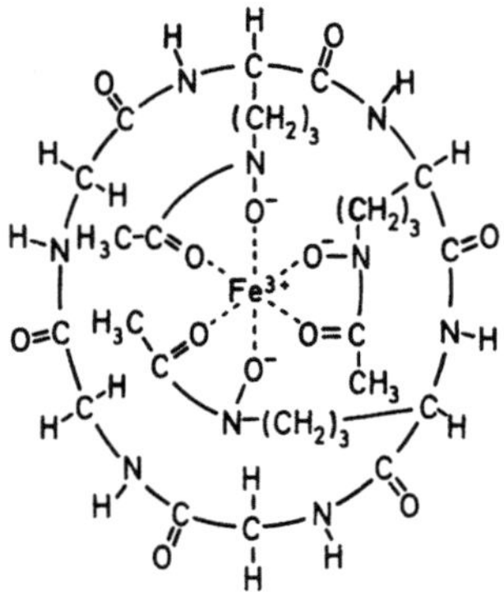

Ferrichrome

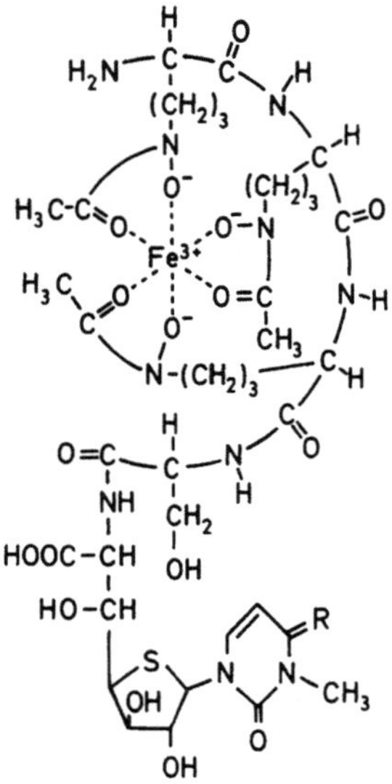

Albomycin δ_1 R= O

Albomycin δ_2 R= N-C(=O)-NH_2

Albomycin ε R= NH

Fig. 1. Structure of ferrichrome and albomycin (see also Fig. 9 for derivatives

of other proteins was strongly reduced. Moreover, the precursor of the FhuA protein accumulated in the cell. Most of its precipitated at the cytoplasmic side of the cytoplasmic membrane and formed large aggregates [20].

The FhuB, FhuC and FhuD proteins have been localized in the cytoplasmic membrane [14,16]. The fact that they are not synthesized as precursors containing signal peptides [16,17] supports their assignment to the cytoplasmic membrane. The amounts formed are so low that they are only detected after radioactive labeling, even when they are expressed on multicopy plasmids. Determination of their primary structure opens several interesting questions. The FhuC protein contains two domains which display strong homology to ATP-binding proteins [17]. The same domains were found in proteins which belong to the class of the so-called binding protein-dependent transport systems [21] through which phosphate, sugars and amino acids enter bacteria. Such binding proteins are located in the periplasmic space, and can be released by treatments which render the outer membrane, but not the cytoplasmic membrane, (temporarily) permeable to macromolecules. Such a treatment is, for example, a sudden change in the osmolarity of the medium, the so-called osmotic shock. However, none of the FhuB, C, D proteins was solubilized, but rather stayed membrane-bound. Only under conditions which preferentially solubilized the cytoplasmic membrane did all 3 proteins become soluble. The following model is proposed. FhuB is an integral membrane protein because it is hydrophobic along its entire length. The FhuC protein binds ATP and thus has to be exposed to the cytoplasm. The only protein left that could function like a binding protein is FhuD. It contains, like FhuC, hydrophobic

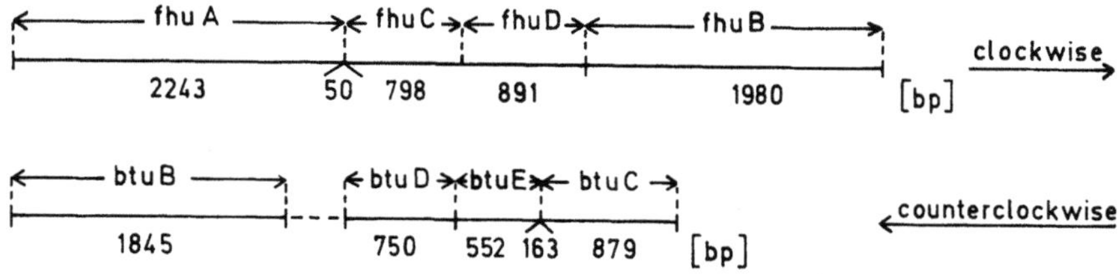

Fig. 2. Comparison of the arrangement of genes involved in ferric-hydroxamate (*fhu*) and vitamin B12 (*btu*) transport. The figures indicate the longest open reading frames of the genes, deduced from the nucleotide sequences including the stop codons. Amino acid sequence data are only available for the FhuA protein and its precursor, supporting the length of the gene. For the *fhuD* gene the translation start site is particularly unclear. [bp] base pairs

and hydrophilic segments [17]. Therefore, part of FhuD could be exposed to the periplasm and serve as the primary binding site for Fe^{3+}-hydroxamate compounds in a way equivalent to the truly soluble periplasmic proteins of binding protein-dependent transport systems.

The Fe^{3+}-hydroxamate transport system displays a remarkable similarity to the transport system for vitamin B12 (Fig. 2). The molecular weight of vitamin B12 (1340 daltons) precludes diffusion through the porin channels of the outer membrane. A receptor protein designated BtuB, the TonB protein, and energy are required in the transport of vitamin B12 across the outer membrane [22,23]. In addition, three genes determine transport proteins, two of which (BtuC, D) were located in the cytoplasmic membrane, and one (BtuE) may be in the periplasm [24]. Even the arrangement of the genes is similar to the genes involved in Fe^{3+}-hydroxamate transport (Fig. 2). However, the genes of both systems are located on the two complementary DNA strands so that the iron transport genes are transcribed clockwise and the vitamin B12 transport genes counterclockwise relative to the genetic map of *E. coli*. The mechanism of transport of Fe^{3+}-hydroxamates and of vitamin B12 seems to be similar because the BtuD protein contains the same ATP-binding consensus sequence as FhuC. Moreover, the FhuB and BtuC proteins are strongly hydrophobic and exhibit some sequence homologies. Apparently, a similar solution was found to transport the large ferrichrome and vitamin B12 complex into cells.

2.3 Fe^{3+}-*Aerobactin Transport*

Aerobactin was detected in 1969 [25]. Ten years later, Williams and Stuart et al. [26,27,28] published evidence for a plasmid-encoded iron transport system which they related to the virulence of *E. coli* strains. Soon it was shown that the siderophore in question was aerobactin [29,30]. Aerobactin consists of two N^6-hydroxyl-N^6-acetyllysine residues linked to citrate (Fig. 3). The iron supply system via aerobactin is usually encoded on large self-transferable, conjugate ColV plasmids. These plasmids determine synthesis of colicin V, a polypeptide that is toxic for certain Enterobacteria. Four contiguous genes encode four enzymes for the biosynthesis of aerobactin [31–33]. These are a N-hydroxylase (oxidase), a N-acetylase and two independently operating proteins which catalyse the attachment of the first and the second N^6-hydroxy-N^6-acetyllysine to citrate [31,32]. In addition to the biosynthesis genes, the plasmide encode

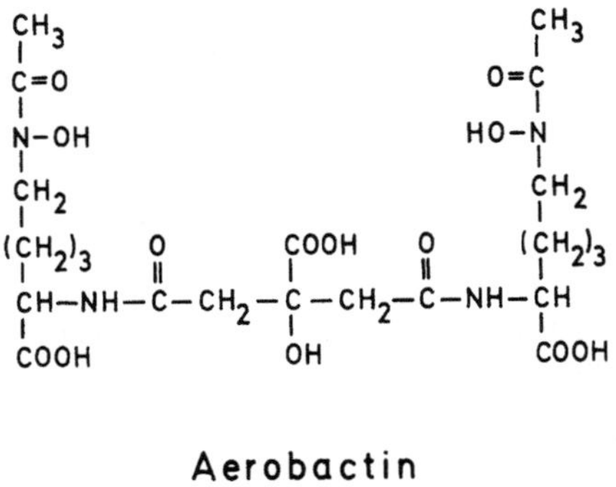

Fig. 3. Structure of aerobactin. It is assumed that Fe^{3+} is bound in an octahedral array by the 4 hydroxamate residues and the central hydroxyl and carboxyl group of citrate

a single transport function, the receptor protein in the outer membrane, which specifically recognizes Fe^{3+}-aerobactin [31-34]. Transport of Fe^{3+}-aerobactin across the cytoplasmic membrane is catalyzed by the *fhuB*, *C*, *D* gene products [35] described in the preceeding section.

Rather different plasmids of *E. coli*, *Salmonella*, *Shigella*, *Klebsiella*, and *Enterobacter* contain a structurally conserved aerobactin region [30-41]. The same aerobactin locus was also found on the chromosome of *E. coli* and *Shigella flexneri* [38,40]. Interestingly, the aerobactin region is flanked on both sides by IS*1* elements in opposite orientation. These could serve to exchange, by transposition or by homologous recombination, the aerobactin region between different DNAs via IS*1* sequences present on different genomes [36,38,39]. In addition, two replication regions were found which may help to maintain the aerobactin system when new plasmids are generated [36]. These data indicate the conservation of the aerobactin system in various enteric bacteria, regardless whether it is encoded on plasmids or on the chromosome. They also suggest mechanisms by which the aerobactin region was distributed among different genomes while maintaining its structure.

2.4 Fe^{3+}-Coprogen Transport

The existence of a specific transport system for coprogen was shown by the identification of an outer membrane protein, termed FhuE, which serves as a constituent of the Fe^{3+}-coprogen uptake system [42]. The same receptor is active in Fe^{3+}-rhodotorulate uptake [42]. Mutants which lack this protein cannot acquire iron via these two siderophores. Additional proteins required for Fe^{3+}-coprogen and -rhodotorulate uptake are the same as for ferrichrome and Fe^{3+}-aerobactin, namely the products of the *fhuB*, *C*, *D*, and the *tonB* genes.

2.5 Fe^{3+}-Dicitrate Transport

The uptake of iron bound to citrate follows the same rules as the other iron transport systems. It requires an outer membrane receptor protein [43], the TonB protein and proteins in the cytoplasmic membrane. Recently, we found that one of the transport proteins (FecB) can be released from the cells by methods which maintain an intact cytoplasmic membrane (L. Zimmermann, unpublished). Therefore, it is possible that a periplasmic component is involved in Fe^{3+}-dicitrate transport as has been discussed for the Fe^{3+}-hydroxamate transport.

A large excess of citrate over Fe^{3+} is necessary to obtain a small homogeneous iron complex which is mainly Fe^{3+}-dicitrate [4]. The molecular weight (434 daltons) of Fe^{3+}-dicitrate should allow access of the compound to the periplasm via the porin channels in the outer membrane. Nevertheless, the FecA outer membrane protein is part of the transport system and only partially dispensable when high concentrations of Fe^{3+}-dicitrate are supplied in the growth medium. Even then, the growth rate of *fecA* mutants with Fe^{3+}-dicitrate as the sole iron source is low.

The most interesting aspect of the Fe^{3+}-dicitrate transport system concerns transcriptional regulation. The transport system of *E. coli* K-12 is induced when 1 mM citrate is present in the growth medium [43,44]. However, citrate is not taken up in measurable quantities by *E. coli* K-12 cells. Apparently citrate induces synthesis of the Fe^{3+}-dicitrate transport system without entering the cytoplasm. This conclusion is supported by additional findings. The iron supply must be low, but some iron is required. In mutants unable to take up citrate-iron, synthesis of the Fe^{3+}-dicitrate system can still be induced [45]. However, mutants devoid of the outer membrane receptor protein, or the TonB function, can no longer be induced [45]. These data can be explained by the assumption that Fe^{3+}-dicitrate has to enter the periplasm, but not the cytoplasm, in order to function as an inducer. Fluorocitrate and phosphocitrate serve as inducers, but transport iron poorly [46]. Metabolic citrate in the cytoplasm is inactive, although in certain mutants it reaches 100 times the concentration required in the growth medium for induction [46]. The threshold concentration of citrate for induction in the medium is 0.1 mM, and the same concentration is required to keep the system induced [46]. This observation also shows that Fe^{3+}-dicitrate is not taken up, since a functional transport system would bring enough Fe^{3+}-dicitrate into the cytoplasm at much lower medium concentrations to maintain the induced state.

Characterization of the system has hitherto revealed 3 genes designated *fecA fecB fecC*, which are arranged in this order on the *E. coli* chromosome. The *fecA* gene encodes the outer membrane receptor protein with a molecular weight of 80,000 daltons, *fecB* determines a periplasmic protein which is synthesized as a precursor (29/27,000 daltons), and *fecC* encodes a protein with an electrophoretic mobility corresponding to a molecular weight of 26,000 daltons (U. Preßler, H. Staudenmaier, L. Zimmermann and V. Braun, unpublished). The 3 genes are transcribed from *fecA* to *fecC*. The major promoter is upstream of *fecA*. Operon fusions between the *lacZ* and the *fecB* genes were constructed to measure transcription rates. Transcription was enhanced 10 fold by supplying citrate to the nutrient broth growth medium, and 15 fold when the available iron was limited by adding dipyridyl at the same time [45]. When iron was completely withdrawn from the medium by addition of deferri-ferrichrome, induction was abolished in a mutant which was unable to take up ferrichrome [46]. On the other hand, transcription of the *fec* locus was also under the control of the iron repressor protein Fur, which does not function as a repressor when no free iron is in the cell [45,47,48]. This supports a regulatory model which predicts that a small amount of iron is required for the formation of extracellular Fe^{3+}-dicitrate and a low intracellular iron concentration thereby preventing repression via the Fur protein. The mechanism of "exogenous induction" exerted by Fe^{3+}-dicitrate remains to be elucidated.

2.6 Fe^{3+}-Enterochelin Transport

The structure of enterochelin, also designated enterobactin, is illustrated in Fig. 4. It is synthesized and released into the growth medium by most Enterobacteriaceae examined to date. Three Fe^{3+}-enterochelin-specific genes have been identified which are involved in transport. The *fepA* gene encodes an outer membrane receptor protein. The molecular weight of the mature protein calculated from the amino acid sequence, deduced from the nucleotide sequence, amounts to 79,908 daltons. FepA is synthesized as a precursor containing a signal peptide consisting of 22 amino acids in addition to the 723 amino acids found in the mature protein [49]. The *fepB* gene determines a periplasmic protein for which molecular weights between 31,500 and 36,500 daltons were determined by gel electrophoresis [50]. The *fepC* gene encodes a membrane-bound 30,500 dalton protein [50].

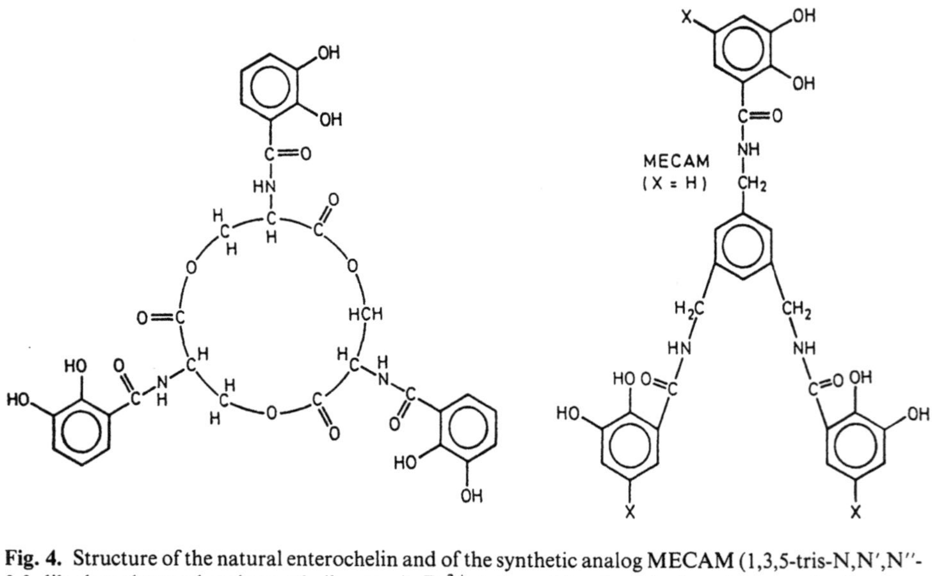

Fig. 4. Structure of the natural enterochelin and of the synthetic analog MECAM (1,3,5-tris-N,N′,N″-2,3-dihydroxybenzoyl-aminomethylbenzene). Fe^{3+} is bound by the 6 hydroxyl groups leading to a triply negative charged molecule [9]

Structural analogs of enterochelin have been synthesized. Fe^{3+}-MECAM (cf:, Fig. 4) supports growth of *E. coli* on iron-limited media [51]. The rate of uptake was similar to that to Fe^{3+}-enterochelin. However, most of it apparently remained unaltered in the periplasm, and delivered little, but enough, iron to the cytoplasm [52].

An esterase determined by the *fes* gene(s) seems to be required for an efficient delivery of Fe^{2+} from Fe^{3+}-enterochelin. Such an esterase cannot hydrolyze MECAM unless it also exhibits an amidase activity. *fes* was mapped within a gene cluster where the *fep* transport genes and the *ent* biosynthesis gene reside at min 13 of the *E. coli* linkage map. The gene order was determined to be *entD fepA fes entF fepC fepB entE B G C A* [53, 54]. Biosynthesis of enterochelin has been previously described [9].

3 Iron Acquisition in the Serum Via Aerobactin and Enterochelin

In the serum and secretory fluids there is virtually no free iron ion due to the extremely strong binding of Fe^{3+} to transferrin (in serum) and lactoferrin (in secretory fluids). Enterobacteria withdraw iron from these binding proteins by secretion of aerobactin and enterochelin. Why did they develop synthetic pathways and transport systems for two siderophores? Biochemically there is no reason for an iron transport system in addition to Fe^{3+}-enterochelin transport. The formation constant of the Fe^{3+}-enterochelin complex is far superior to any other Fe^{3+}-siderophore including Fe^{3+}-aerobactin. In addition, the transport rate of Fe^{3+}-enterochelin into *E. coli* compares favorably to that of Fe^{3+}-aerobactin. However, several factors seem to contribute to the preferential use of Fe^{3+}-aerobactin in serum. Upon delivery of iron to cells, aerobactin is not destroyed as is enterochelin. Aerobactin can repeatedly transport iron [55], whereas one molecule of enterochelin seems to carry only one iron ion and is then hydrolyzed to products which are poor iron carriers, and also do not serve as precursors of enterochelin synthesis [9]. In addition, enterochelin-specific antibodies have been found in normal human serum at concentrations which inhibited the growth of *Salmonella typhimurium* in complement-inactivated serum, unless the cells produced enough enterochelin to neutralize the antibodies [56]. The antibodies are of the IgA class and inhibit iron uptake into *E. coli* via enterochelin but not via ferrichrome and citrate [57]. A second class of antibodies directed against the lipopolysaccharide also impairs iron uptake. Inhibition of enterochelin synthesis, or release, was proposed to explain these data [58, 59]. We have recently found that human isolates of *E. coli* producing enterochelin are more sensitive to the bactericidal action of human serum than derivatives which have lost the ability to synthesize enterochelin (H. Chehadeh, H. Rotering, V. Braun, unpublished data). Enterochelin in contrast to aerobactin also adsorbs to serum albumin [60]. At a siderophore concentration of 0.3 mM, which was 10 times higher than Fe^{3+}-transferrin, aerobactin transported iron into cells of *E. coli* suspended in serum, whereas enterochelin was virtually inactive [61]. In a synthetic minimal medium, addition of 5 nM aerobactin supported growth at maximum rate of an *E. coli* strain unable to synthesize, but able to transport, Fe^{3+}-aerobactin and Fe^{3+}-enterochelin while 500 nM enterochelin had to be supplied for optimum growth [62]. Addition of serum (10% vol/vol) inhibited enterochelin, but not aerobactin-promoted growth. Transfer of ColV plasmids, bearing the Fe^{3+}-aerobactin system, rendered *E. coli* K-12 strains capable of rapid growth in undiluted calf serum. By employing appropriate mutants, growth stimulation was clearly related to iron acquisition. Strains containing the Fe^{3+}-aerobactin system grew from densities of 10^3 to 10^8 cells per ml within 24 h, while aerobactin deficient strains, regardless whether they expressed the enterochelin system, ceased growth below 10^5 cells per ml and died [55].

The frequency of occurrence of the aerobactin iron supply system has been studied in isolates of *E. coli* obtained from human and animal sources. In clinical isolates of *E. coli* aerobactin was produced by 45% from a total of 96 strains isolated from blood, and by 41% from a total of 61 faecal isolates [62]. For comparison, the incidence for hemolysin synthesis and serum resistance was 37% and 28% in blood isolates, and 75% and 43% in fecal isolates, respectively. Similar figures were found in other studies. The frequency of aerobactin synthesis among *E. coli* strains associated with disease, in contrast to commensal strains, does not seem to be so much different as to attribute the Fe^{3+}-aerobactin transport system a decisive role in virulence. But there is sufficient evidence to suggest that this iron transport system contributes to the outcome of an infection [6]. The relative importance of the aerobactin system certainly depends on the combination with other virulence factors. For example, a high incidence of aerobactin synthesis was found in fimbriated urinary tract isolates of *E. coli* [63,64], suggesting an important role of iron supply via aerobactin for the colonisation of the urinary tract. In fact, iron receptor proteins of the *E. coli* outer membrane were found derepressed, when examined without further subculturing, from patients with urinary tract infections [65]. Epidemiological studies alone will not provide a conclusive answer under what circumstances the Fe^{3+}-aerobactin system serves as a virulence factor. They must be completed by studies employing animal infection models, in which various combinations of virulence factors are assembled by genetic means in a particular strain [66]. Studies of this kind will have to include the other iron transport systems. For example, the nearly ubiquitous Fe^{3+}-enterochelin system may provide the basic iron supply so that the Fe^{3+}-aerobactin system may only be required under certain conditions. One also has to bear in mind the frequently increased synthesis of toxins under iron-limiting conditions. An ample iron delivery may counteract the action of other virulence factors such as toxins.

4 Iron Transport Systems in Bacteria Excluding E. Coli and Closely Related Species

Numerous reports provide evidence for the presence of iron transport systems in bacteria. In Gram-negative bacteria, outer membrane proteins were identified which

Table 1. Outer membrane proteins of gram-negative bacteria derepressed in low-iron medium

Organism	Size of the proteins in kilodaltons	Refs.
Escherichia coli	83, 81, 80, 78, 76, 74.5[a], 74	82)
Salmonella typhimurium	83, 81, 74	83)
Shigella flexneri	83, 81, 74	84)
Klebsiella pneumoniae	66, 69, 70, 73, 75, 78, 83	87)
Vibrio cholerae	220[b], 77, 76, 75, 73, 62	85)
Yersinia species	81, 79, (89, 70, 68, 27.5)[c]	86)
Proteus species	69–83	87)
Neisseria species	97, 86, 76, 37	68, 88)
Pseudomonas species	80, 75, 14	89, 90)
Aeromonas salmonicida	83, 78, 77	91)
Vibrio anguillarum	86[a], 79	92)
Azotobacter vinelandii	93, 85, 77	93)
Vibrio vulnificus	81, 74	94)
Haemophilus species	70, 73, 74, 76, 86	95)

[a] Proteins determined by plasmids
[b] Apparently a protein at the cell surface loosely associated with the outer membrane
[c] Two additional high-molecular-weight proteins were observed

were strongly expressed in low-iron media (Table 1). Most of them displayed electrophoretic mobilities similar to the proteins of *E. coli* for which an iron-siderophore transport function had been demonstrated. Such proteins were shown to be present in *Neisseria* species, which in addition to the proteins of high molecular weight expressed iron-regulated proteins with molecular weights of 37,000 and 19,800 [67,68]. Apparently, *Neisseria* do not secrete siderophores. Nevertheless, they are able to withdraw iron from transferrin, lactoferrin and heme proteins [69,70,71]. *Pseudomonas* species produce outer membrane proteins and a number of siderophores. Their structures are shown in Figs. 5–7. They are designated pyochelin, pseudobactin and pyoverdin. Genetic studies are under way to determine the pathways of *Pseudomonas* siderophore production and regulation [73–75]. Table 1 provides a survey of genera in which outer membrane proteins regulated by iron have been determined.

Distinct mechanisms of iron acquisition exist within the genus of mycobacteria. Four systems were found: A high affinity system active at low concentrations of iron is energy-coupled, and iron is transported as ferriexochelin, which designates a group

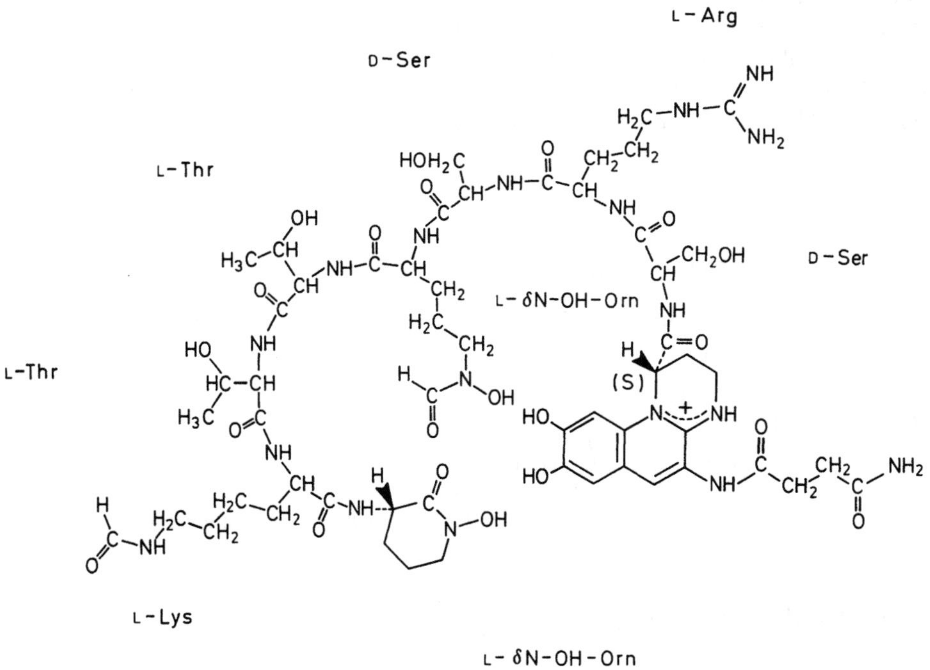

Fig. 5. Structure of pyoverdin Pa of *Pseudomonas aeruginosa* [101]

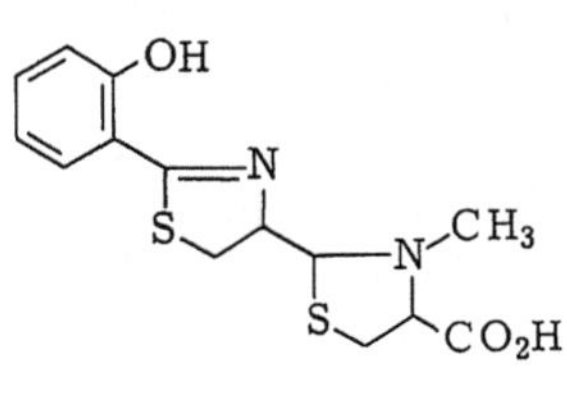

Fig. 6. Structure of pyochelin

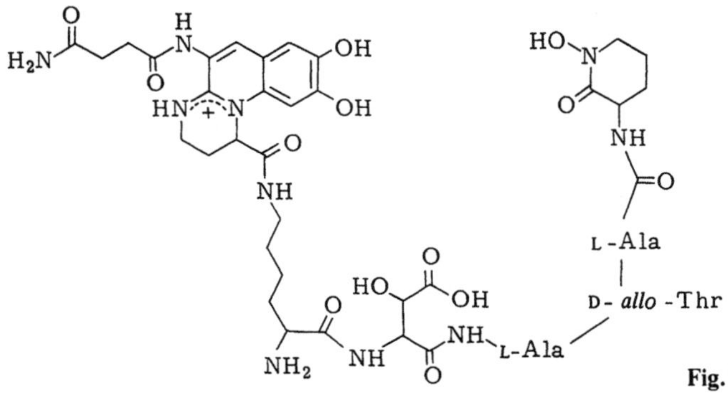

Fig. 7. Structure of pseudobactin

of extracellular siderophores of unknown structure. At higher iron concentrations (above 10 μM) ferriexochelin diffuses into cells at sufficient rates to support growth. Another system uses salicylate, which is released by mycobacteria under conditions of iron-limited growth. However, its role in iron supply is unclear. The third transport process operates via ferric dicitrate. In addition, Mycobacteria contain large amounts (up to 10% of the cell dry weight) of a lipid-soluble iron binding compound designated mycobactin, which seems to function as an iron storage rather than a transport compound [79,80]. The common structure of mycobactins is shown in Fig. 8. [81].

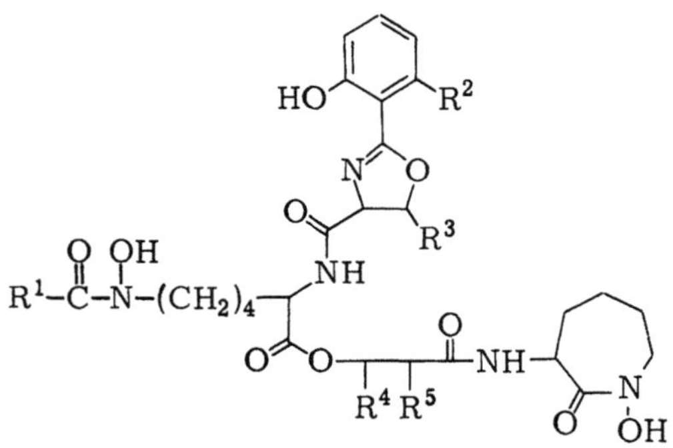

Fig. 8. Structure of myobactin

5 Siderophores of Fungi

5.1 General Features

Like almost all aerobic and facultative anaerobic organisms fungi also depend on a continuous iron supply during the active growth phase. Part of the assimilated iron is required for the synthesis of enzymes of the respiratory chain in the mitochondria. Another part of the assimilated iron is required for enzymes catalyzing cellular redox reactions. There is also evidence that a certain amount of cellular iron in fungal cells is present as a kind of iron store which, however, is not generally ferritin [96]. While the individual iron containing components of fungal cells may be extracted and analyzed in vitro, the process of iron assimilation in fungi can only be studied with intact cells. Furthermore, since mutants of iron transport in fungi could not be obtained until now, the transport systems of fungi have been characterized predominantly by kinetic studies. The assimilation of iron requires the development of specific iron transport systems within the cytoplasmic membrane and its coupling to the energy metabolism of the cell. Recent results have shown that iron uptake among fungi is not dependent on a single iron uptake mechanism. It rather appeares that the most essential element for life has more than one entry into the fungal cells [97]. The studies on iron uptake, in particular the transport measurements with radioactive iron (^{55}Fe or ^{59}Fe) indicate that iron may be taken up either in the ferrous or in the ferric form. Whereas uptake of ferrous ion is still not well understood, uptake of ferric ion has been analyzed in more detail. A prerequisite for ferric ion uptake into the fungal cell is its solubilization either from polymeric iron hydroxides or from other iron containing complexes. For that purpose most fungi synthesize Fe(III)-specific ligands, termed siderophores. In this chapter we therefore discuss mainly the transport of ferric ion mediated by siderophores and describe the structures which have been elucidated so far and examine the specificity of siderophore mediated iron uptake in *Neurospora crassa* and some species of the genera *Ustilago*, *Rhodotorula*, *Fusarium*, *Aspergillus*, *Penicillium*. Several reviews are available on these topics [72, 75, 98, 99, 100, 101].

5.2 Biosynthesis

The production of siderophores is triggered when the iron content of the medium decreases below a value of 1 μM, which is about ten times higher than the amount of iron which triggers siderophore production in bacteria. The reason for this is unknown but it may be assumed that in eukaryotic cells iron has to overcome several compartmental barriers, so that the the internal iron content responsible for triggering siderophore biosynthesis may be finally identical in fungi and bacteria. Another explanation for the earlier onset of siderohore excretion in fungi may be that larger amounts of iron have to be provided before growth can continue. Thus, the definition of microbial siderophores includes that its biosynthesis is regulated by the external and finally by the internal iron content. Although the genetic regulation of fungal siderophore production is still unresolved, there is every reason to assume that the regulation of siderophore biosynthesis proceeds by an iron-containing repressor at the DNA level.

5.3 Isolation

Hydroxamate siderophores can be obtained in high amounts by cultivation of the producing strain in a chemically defined medium without added iron. Although the initial iron content of such a medium is not low enough to immediately induce siderophore production, the medium is gradually depleted of iron during further growth of the organism. The maximum of siderophore production has to be determined for the individual strain, but for most fungi a cultivation time of 5 days is sufficient for a complete derepression of siderophore biosynthesis. Since oxygen is required for the hydroxylation of the δ-N of ornithine by an δ-N-oxidase, the culture has to be aerated for optimal siderophore production. A common procedure for the isolation of siderophores from a culture is to add ferrous sulfate to the culture filtrate under stirring until the typical brown-orange color does not increase further and passing the culture filtrate through an Amberlite XAD-2 column [102]. After washing with water the siderophore fraction can be desorbed by methanol or acetone-water 1:1. The crude siderophores can be purified using gel filtration procedures and siliga gel chromatography. HPLC on reversed phase columns (RP-8, RP-18) as well as thin layer chromatography on silica gel plates with chloroform-methanol-water mixtures as solvent systems are used for separation and identification of the various siderophores [129].

6 Ferrichromes

Ornithine is the most important constituent in all fungal siderophores. It seems reasonable that a non-proteinogenic amino acid is used for the biosynthesis of siderophores in order not to vaste essential components required for cell growth. Ornithine is hydroxylated and acylated at the δ-N atom yielding the δ-N-hydroxy-δ-N-acyl-ornithine residue. Three of these bidentates constitute the iron binding center of the fungal trihydroxamate siderophores. All ferric trihydroxamates are octahedral complexes in which the coordinated ferric iron is d^5, high spin and rapidly exchange-

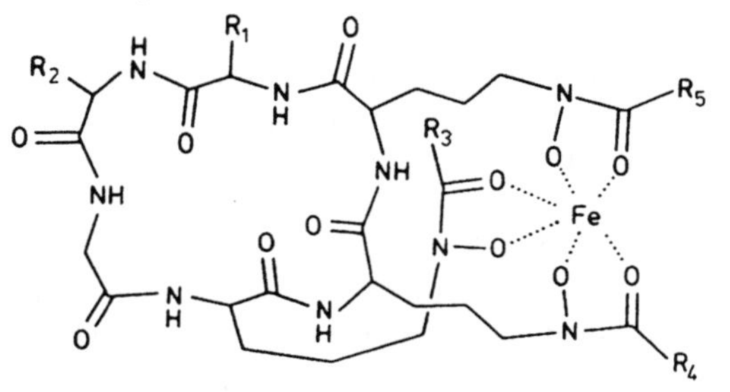

Siderophore	R_1	R_2	R_3	R_4	R_5
ferrichrome	H	H	Ac	Ac	Ac
ferricrocin	H	$-CH_2-OH$	Ac	Ac	Ac
ferrichrysin	$-CH_2-OH$	$-CH_2-OH$	Ac	Ac	Ac
ferrirhodin	$-CH_2-OH$	$-CH_2-OH$	cis-AM	cis-AM	cis-AM
ferrirubin	$-CH_2-OH$	$-CH_2-OH$	trans-AM	trans-AM	trans-AM
asperchrome B1	$-CH_2-OH$	$-CH_2-OH$	Ac	trans-AM	trans-AM
asperchrome D1	$-CH_2-OH$	$-CH_2-OH$	trans-AM	Ac	Ac

Ac = acetyl, AM = anhydromevalonyl.

Fig. 9. Structures of ferrichrome-type siderophores (Ac, AM = hydroxamic acid residues)

able. Ferrichrome is regarded as the type-siderophore of a series of cyclic peptide siderophores collectively termed ferrichromes (Fig. 9). It was first detected in cultures of the smut fungus *Ustilago sphaerogena* [106] but was later found in a variety of other genera such as *Aspergillus, Penicillium, Cintractia, Farysia, Moesziomyces, Schizonella, Sphacelotheca, Sporisorium, Graphiola, Protomyces* and *Tilletiaria* [72]. Ferrichrome contains a cyclic gly_1-gly_2-gly_3-orn_1-orn_2-orn_3-hexapeptide sequence in which the ornithine residues are δ-N-hydroxylated and δ-N-acetylated. Whereas -gly_3- is a necessary constituent in all ferrichromes for the development of a β-II turn, -gly_2- may be substituted by -ser-: ferricrocin [107,108] or by -ala-: ferrichrome C [109]. In the case that both, -gly_2- and -gly_1- are substituted by -ser- the compound is named ferrichrysin [107,108]. Ferrichrysin itself is the parent compound of a further series of ferrichrome-type compounds, in which the acyl residue may be either methylglutaconic acid: ferrichrome A [110], trans-anhydromevalonic acid: ferrirubin [111], cis-anhydromevalonic acid: ferrirhodin [111] or a mixture of acetyl and trans-anhydromevalonyl residues: asperchromes [112].

6.1 Fe-Rhodotorulic Acid

Rhodotorulic acid is the dimer of δ-N-hydroxy-δ-N-acetyl-L-ornithine forming a diketopiperazine ring (Fig. 10). It has been found in low-iron cultures of the fungus *Rhodotorula pilimanae* [113,114] and other heterobasidiomycetes yeasts [72,115,116]. The stability constant is log $\beta_{110} = 31.2$ and pM = 21.9 [5] indicating that the dihydroxamic siderophore rhodotorulic acid is a fairly good chelator. At neutral pH it forms a $Fe_2(RA)_3$ complex which is believed to adopt a triply bridged binucleic complex [117,118]. Uptake studies revealed that *Rhodotorula pilimanae* assimilated iron from labeled Fe-rhodotorulate but denied iron from ferrioxamine B, ferricrocin or coprogen [119,120].

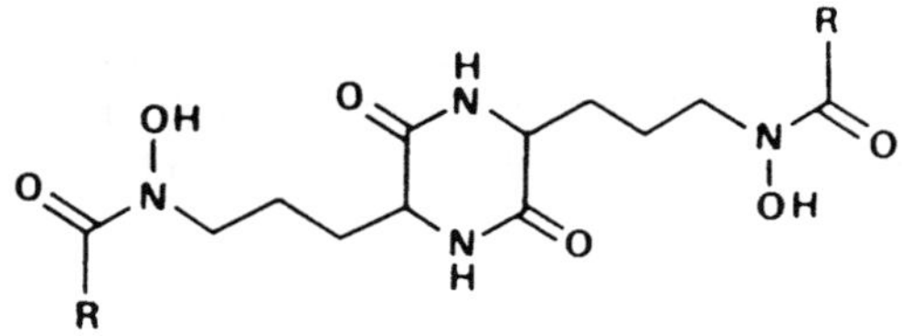

Fig. 10. Rhodotorulic acid, R = CH_3; dimerum acid, R = *trans*-5-hydroxy-3-methyl-pent-2-enyl (= *trans*-anhydromevalonyl = *trans*-AM)

6.2 Fusigens

Iron complexes composed of δ-N-hydroxy-δ-N-*cis*-anhydromevalonyl-L-ornithine residues (Fig. 11) are called fusarinins [121] or fusigens [122]. Fusigens are produced by strains of *Fusarium, Penicillium* and *Aspergillus.* Although linear fusarinin complexes (fusarinin, fusarinin A, fusarinin B) have been detected in culture filtrates they probably originate from the degradation of the cyclic triester fusarinin C (fusigen). Some strains of *Aspergillus* [123] and *Mycelia sterilia* [124] also synthesize the more stable and lipophilic triacetylfusigen.

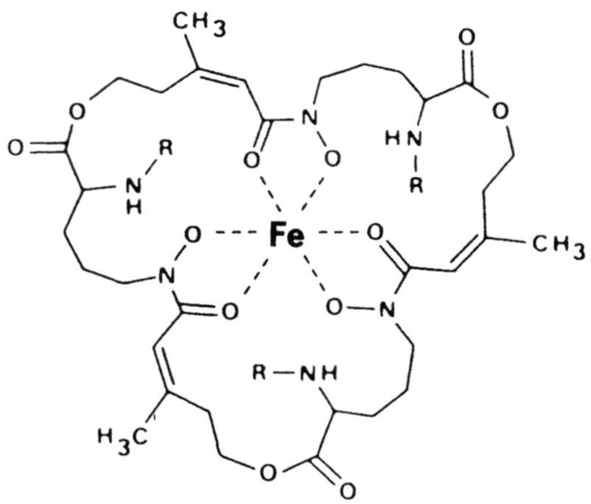

Fig. 11. Fusigen (= fusarinin C), R = H; triacetylfusigen, R = acetyl

6.3 Dimerum Acid and Coprogens

Two residues of δ-N-hydroxy-δ-N-*trans*-anhydromevalonyl-L-ornithine residues (*trans*-fusarinine) connected head to head in a diketopiperzine ring are called dimerum acid (Fig. 10) [125]. Since dimerum acid appeared later than coprogen during fermentation, it may probably represent a degradation product of coprogen (Fig. 12) which also contains one molecule of dimerum acid esterified to a further α-N-acetyl-L-δ-N-hydroxy-δ-N-*trans*-anhydromevalonyl-L-ornithine residue. Coprogen is the main siderophore of *Neurospora crassa* [126] and some strains of *Penicillium* [72]. Besides desacetylcoprogen (coprogen B), produced by strains of *Fusarium* [125], several other coprogen analogues have been isolated, e.g. triornicin and isotriornicin from low-iron culture of *Epicoccum purpurascens* [127,128] and neocoprogen I and neocoprogen II from low-iron cultures of *Curvularia subulata* [129]. Triornicine is structurally similar to coprogen, exept that the esterlinked ornithine contains an δ-N-acetyl group. Isotriornicin and neocoprogen I are found to be identical, both possessing an δ-N-acetyl group at the diketopiperazine end and a δ-N-*trans*-anhydromevalonyl group at the ester linked ornithine. Neocoprogen II contains acetyl residues at both ends of the molecule.

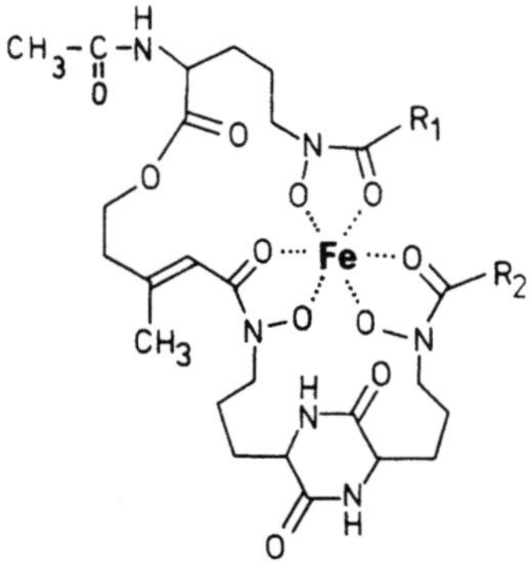

Fig. 12. Coprogen: $R_1 = R_2 =$ *trans*-5-hydroxy-3-methyl-pent-2-enoyl (= *trans*-anhydromevalonyl = *trans*-AM); neocoprogen I (= isotriornicin), $R_1 =$ *trans*-AM, $R_2 = CH_3$(acetyl); neocoprogen II, $R_1 = R_2 = CH_3$(acetyl)

7 Physico-Chemical Properties of Siderophores

7.1 Formation Constants and pM Values

All siderophores of fungal origin contain three hydroxamate bidentate ligands which represent fairly weak acids with pKa's at about 9:

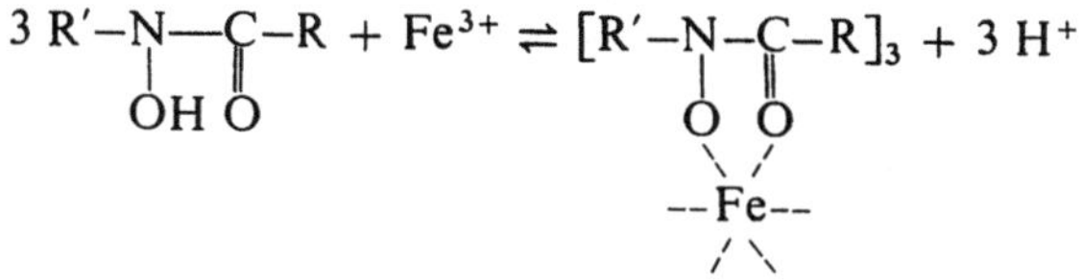

Depending on the prevailing pH, di-, tetra- and hexadentate complexes are possible yielding 1:1, 1:2 and 1:3 iron-bidentate complexes. The predominant form in neutral and slightly acidic solution (pH 7–2) is the hexadentate octahedral iron(III) chelate possessing no net charge. Trihydroxamate chelates show formation constants in the range of 10^{28}–10^{32} [102]. Since formation constants have no direct relevance to physiological conditions, pM-values have been introduced, which allow a direct comparison of the effectiveness of different chelators. The pM-value is the negative logarithm of the free metal ion concentration (hexaaquo Fe^{3+}) in equilibrium with the iron complex where the total Fe^{3+} concentration is 10^{-6} M and the total ligand concentration is 10^{-5} M and pH is 7.4. Thus, under these defined conditions a direct comparison of different siderophores is possible. For example the pM of enterobactin is 35.6, indicating that under the given conditions the free iron concentration is $10^{-35.6}$ instead of 10^{-52} as indicated by the formation constant. On the other hand the pM values of the fungal hydroxamates (Table 2) are in the range of 21.9–27.5, which are almost in the same order of magnitude as the formation constants. Bacterial catecholates show higher complex stabilities at alkaline pH values [103]. The stability constant for enterobactin ($K \sim 10^{52}$) for example significantly decreases in acid solution. Contrary to the bacterial catecholate-type siderophores, the fungal hydroxamate-type siderophores are optimally designed for binding and transport of ferric ions in a slightly acidic environment.

Table 2. Physico-chemical constants of fungal hydroxamate siderophores[a]

Siderophore	pK_1 [b]	β_{110} [c]	pM[d]	$E_{1/2}$ mV vs NHE
Ferrichrome	9.83	29.1	25.2	—
Ferricrocin	9.92	30.4	26.5	—412
Ferrichrysin	—	30.0	25.8	—
Ferrichrome A	—	32.0	—	—448
Rhodotorulic acid	9.44	(31.2)	21.9	—359
Coprogen	9.16	30.2	27.5	—447

[a] Values are taken from Ref. [5] and [102]
[b] $K_1 = [HL^{-2}]/[H^+][L^{-3}]$;
[c] $\beta_{110} = [FeL^{3-n}]/[Fe^{3+}][L^{n-}]$;
[d] $pM = -\log[Fe(H_2O)_6^{3+}]$ when the total $[Fe] = 10^{-6}$ M, $[L] = 10^{-5}$ M, pH = 7.4

7.2 Chemical Stability

Regarding the chemical stability of fungal siderophores, it is well known that the cyclic hexapeptide siderophores are resistant to the attack of known proteinases, whereas the ester type fusarinins are more amenable to enzymatic degradation [104]. It is interesting to note that ester- and peptide-type siderophores are often both produced by the same fungal strain. Fungal strains have the ability to produce a variety of structurally different siderophores, which may point to a selective advantage of certain siderophores under particular environmental conditions. Because of the extraordinary resistance of the cyclic peptide siderophores (ferrichromes) against degradation, ferrichromes may be found in earth in appreciable amounts [105] and may thus serve as iron sources for a variety of saprophytic microbes.

7.3 Conformation and Configuration

Crystallographic data revealed that the ferrichrome molecule possesses a fairly rigid structure [120]. NMR data and tritium exchange experiments proved that the solution structure is in good agreement with the crystal structure [131, 132, 133]. The conformation of the peptide backbone is quite similar in all the ferrichromes, e.g. all ferrichromes form a transannular H-bond associated with a β(II) bend within the peptide backbone [134].

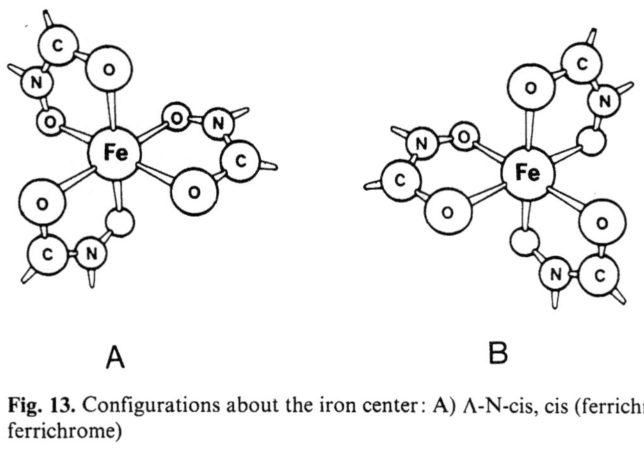

Fig. 13. Configurations about the iron center: A) Λ-N-cis, cis (ferrichrome), B) Δ-N-cis, cis (enantio-ferrichrome)

The absolute configuration of the iron coordination is Λ-cis in all ferrichromes (Fig. 13A) as determined x-ray diffraction in crystals and by circular dichroism spectra in solution [134, 102]. This uniformity in the iron coordination center obviously results from the chirality of the peptide backbone which is composed of L-ornithine residues. The synthetic enantio-ferrichrome [135] and enantio-ferricrocin [136], adopt a Δ-configuration about the iron center (Fig. 13B) as shown by CD-spectroscopy.

8 Transport of Siderophores

8.1 Structure and Transport

A comparison of the transport properties of natural ferrichromes and enantio-ferrichromes in fungi has shown that the fungal siderophore transport system does not recognize the enantiomeric ferrichromes [136, 137]. These results indicated that the stereospecificity of the chelate molecules is an important feature for the recognition and transport process in fungi. Although in *E. coli* the natural ferrichrome was the preferred transport molecule, enantio-ferrichrome was still transported with about half the rate of ferrichrome. This may be explained by a less precise recognition of the enterobacterial FhuA receptor, or by the existence of additional uptake routes for iron from enantioferrichrome [138]. The inability to take up enantioferrichromes in fungi indicates quite clearly the existence of a membrane component recognizing structure and configuration of the siderophores. This recognition site is obviously part of a siderophore transport system within the cytoplasmic membrane. This also excludes the possibility that siderophores permeate the fungal cytoplasmic membrane by a diffusion controlled process, as it is the case for the classical ionophores [139].

Synthetic analogues are of great value for elucidating the specificity of the siderophore transport. While the enantiomeric ferrichromes have clearly shown a highly stereoselective transport across the cytoplasmic membrane, the use of several naturally occuring ferrichromes demonstrated the importance of the δ-N-acyl groups around the iron binding center. Thus, ferrichrome A and ferrirubin, both possessing branched N-acyl residues, were not transported by the ferrichrome transport systems in *Neurospora crassa* [140] and *Ustilago sphaerogena* [141]. Moreover, a new synthetic analogue of ferrichrome, retrohydroxamate ferrichrome has been reported to be biologically as active as the natural compound, indicating that reversing the three —NOH—CO—CH_3 groups to —CO—NOH—CH_3 yielded a biomimetic analogue [142]. In contrast, the total lack of biological activity of desmethylretrohydroxamate (—CO—NOH—H) confirmed the importance of the methyl groups at the chelation center for cellular recognition [143]. Recent results have shown that the siderophore transport system *in Neurospora crassa* is very sensitive to variations of the length, kind and number of the N-acyl residues of both the ferrichromes and coprogens [144]. Thus, in the coprogen series uptake decreased in the order of coprogen, neocoprogen I and neocoprogen II, indicating that gradual replacement of the N-*trans*-anhydromevalonyl (AM) groups by N-acetyl (AC) groups had an adverse effect on uptake. The reverse effect was observed in the ferrichrome series where uptake decreased in the order ferrichrysin (3 AC), asperchrome D1 (2 AC, 1 AM), asperchrome B1 (1 AC, 2 AM), and ferrirubin (3 AM). Conformation of the anhydromevalonyl group (*cis* or *trans*) of ferrichromes was also an important determinant in the recognition process. Ferrirhodin contains three *cis*-AM groups instead of the three *trans*-AM groups found in ferrirubin. Uptake rates by these two compounds were quite different (Table 3), indicating that the N-acyl conformation of the anhydromevalonyl groups is also crucial for uptake. A variety of changes in the peptide backbone have been shown to exert little or no effect on uptake. Thus the heptapeptide siderophore, tetraglycylferrichrome [145], containing a -$(Gly)_4$-sequence, revealed nearly identical transport rates as compared with ferri-

Table 3. Transport rates of siderophores in *Neurospora crassa* (*arg-5 ota aga*)[a]

Siderophore	Uptake (p moles mg^{-1} min^{-1})
Ferrichrome[b]	105
Ferricrocin[b]	95
Ferrichrysin	70.6
Asperchrome D1	39.3
Asperchriome B1	6.66
Ferrirubin	4.66
Ferrirhodin	68
DDF[c]	2.46
CDDF[d]	2.73
Coprogen	80
Neocoprogen I	63.3
Neocoprogen II	·30.6

[a] Data taken from Ref. [144];
[b] data taken from Ref. [136];
[c] DDF = des(diserylglycyl)ferrirhodin [146];
[d] CDDF = capped DDF = $^{\alpha}$N-acetyl-des(diserylglycyl)ferrirhodin methyl ester

chrome, containing a -$(Gly)_3$-sequence. Furthermore, the K_m values ($\sim$5 μM) for uptake of ferrichrome, ferricrocin and ferrichrysin were quite similar although the serine content of the cyclic hexapeptide differs. The semisynthetic derivatives of ferricrocin, O-(phenylyl-carbamoyl)-ferricrocin and O-(sulfanilyl-carbamoyl)-fearicrocin showed only slightly decreased uptake rates [136]. Summarizing the transport measurements with natural and synthetic N-acyl derivatives of siderophores we may conclude that the iron center and its surrounding functionalities are the predominantly recognized part of the siderophore molecule. Although it is clear from the previous discussion that the peptide backbone of ferrichromes is not directly involved in the recognition process we have evidence that the cyclic nature of the peptide is an absolute requirement for the transport process. The function of the cyclic peptide portion of the ferrichromes and asperchromes can be most directly investigated from the results on ferrirhodin and des(diserylglycyl)ferrirhodin (DDF). In DDF three ornithyl residues (with three *cis*-AM groups) form a tripeptide which is identical to the ornithyl part of ferrirhodin [146]. The iron coordination geometry of DDF, determined by circular dichroism spectroscopy, was found to be Λ-cis as seen in all ferrichromes. While ferrirhodin was taken up by *Neurospora crassa* with rates similar to ferrichrysin, DDF was not taken up at all, although it was bound to the recognition sites of the transport system as indicated by a 74% inhibition of coprogen uptake. To eliminate the dipolar ion in DDF, N-acetyl DDF and N-acetyl DDF-methylester (CDDF) were prepared and found to be similarly inactive. These results clearly indicate that the peptide portion of the ferrichrome molecule is important for the function. The observation that part of the molecule being transported is responsible for binding, whereas another part is responsible for transport across the cytoplasmic membrane, seems to be a general feature of biological phenomena involving the transport of small molecules by macromolecules.

8.2 Recognition Sites

The suggestion of separate recognition sites for coprogens and ferrichromes in *Neurospora crassa* resulted from the different transport behavior of the series coprogen, neocoprogen I, neocoprogen II, and the series ferrichrysin, asperchrome D1, asperchrome B1, ferrirubin, in which the number of N-trans-anhydromevalonyl groups had an adverse effect on uptake. Further evidence for distinct recognition sites came from studies with partially resolved coordination isomers of chromic desferricoprogen [147]. Crystallographic data and circular dichroism spectra revealed that the absolute configuration around the metal center of ferrichromes and coprogens is quite different. Thus, the configuration in the ferrichrome series is Λ-cis, while the configuration of the coprogens in solution is predominantly Δ-trans (Fig. 14). Neither fraction of the isolated chrom-substituted coprogens had a significant effect on the uptake of ferricrocin. However, the Δ-trans isomer fraction of chrom-substituted coprogen significantly reduced the uptake of ^{55}Fe-labelled coprogen. These results are strong evidence for the existence of separate siderophore recognition sites in fungi. Additional evidence came from uptake studies with *Penicillium parvum*, which is able to produce ferrichrome but is unable to produce coprogen. Uptake studies revealed that while ferrichrome is well taken up, coprogen was not, indicating that in this fungus a coprogen recognition site is lacking [136].

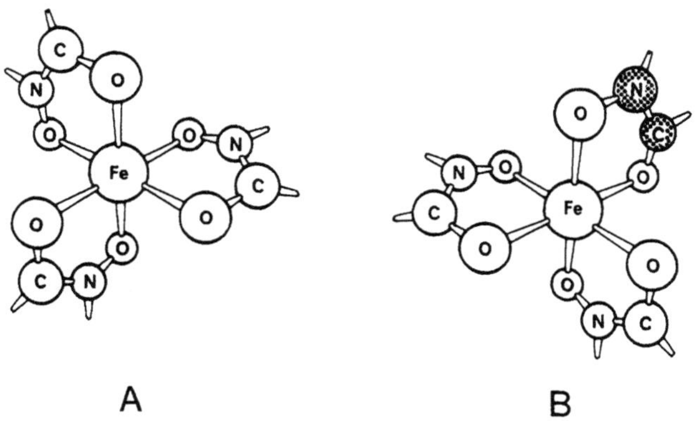

Fig. 14. Configurations about the iron center: A) Λ-N-cis, cis (ferrichrome), B) Δ-C-trans, trans (coprogen)

8.3 A Common Siderophore Transport System

The inhibition studies with labeled coprogen (Table 4) pointed to some commonality during the transport of siderophores in *Neurospora crassa*. *Neurospora crassa* produces ferricrocin [148] and coprogen which show competitive inhibition during uptake. This led us to conclude that ferrichromes and coprogens were both transported by a shared transport system (Fig. 15) in the cytoplasmic membrane [136]. The stereospecificity of the uptake suggested separate recognition sites for the different siderophore types, although a strong allosteric influence of both sites remains an alternative explanation [136].

Table 4. Inhibition of ^{14}C-coprogen uptake in the presence of ferrichrome-type and coprogen-type siderophores[a]

Siderophore	Kind and number of iron-surrounding N-acyl residues	Inhibition of coprogen uptake at equimolar concentration (10 μM)
Ferrirubin	3 trans-AM, 0 AC	97%
Asperchrome B1	2 trans-AM, 1 AC	96%
Asperchrome D1	1 trans-AM, 2 AC	92%
Ferrichrysin	0 trans-AM, 3 AC	57%
Ferrirhodin	3 cis-AM, 0 AC	74%
CDDF[b]	3 cis-AM, 0 AC	72%
N-acetyl DDF	3 cis-AM, 0 AC	50%
DDF[c]	3 cis-AM, 0 AC	17%
Coprogen	2 trans-AM, 0 AC	50%
Neocoprogen I	1 trans-AM, 1 AC	37%
Neocoprogen II	0 trans-AM, 2 AC	21%

[a] Data taken from Ref. [144];
[b] CDDF = N-acetyl-des(diserylclycyl)ferrirhodin;
[c] DDF = des(diserylglycyl)ferrirhodin

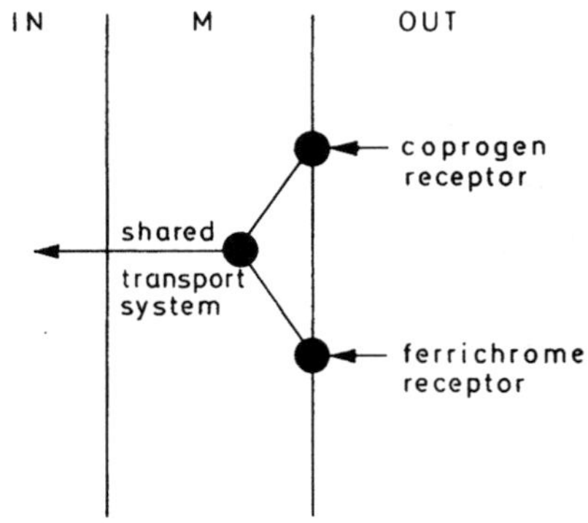

Fig. 15. Schematic illustration of the two recognition sites and the common transport system for siderophores in the cytoplasmic membrane of *Neurospora crassa* (taken from Ref. [100])

8.4 Mechanisms of Transport

Three different mechanisms of siderophore transport in fungi have been reported: i. transport of the intact siderophore into the interior of the fungal cell, as found in *Neurospora crassa* [149]. ii. transport of the siderophore to the cell surface (taxi mechanism) and release of the ferric ion with subsequent assimilation without reduction. This mechanism has been observed with Fe-rhodotorulate in *Rhodotorula pilimanae* [119, 150]. iii. transport of the siderophore to the membrane with subsequent reduction and uptake of the ferrous ion. This reductive taxi mechanism has also been shown to occur with ferrichrome A in *Ustilago sphaerogena* [151, 152]. Although all these mechanisms require recognition and binding of siderophores to a particular site on the cytoplasmic membrane, they differ significantly in place, time and mode of iron release. In *Neurospora crassa* the transport of siderophores leads to an accumulation of siderophores (coprogen or ferrichromes) inside the cells. Reductive removal of iron from siderophores proceeds very slowly as determined by enzymatic assays [149] as well as by Mössbauer spectroscopy [153]. In *Rhodotorula pilimanae* the ferric ion is possibly removed from the chelate molecule outside the cell and requires no reduction mechanism, as the non-reducible gallium-rhodotorulate complex was also transported by this fungus [150]. In *Ustilago sphaerogena* two mechanisms are operative, one that transports ferrichrome as a whole into the interior of the cell and another one that takes up iron from ferrichrome A by means of a reduction step at the membrane [152]. The actual mechanism of reduction of siderophores on the cytoplasmic membrane is still unclear but it may be speculated that fungi possess membrane bound electron transport enzymes as has been suggested in plant iron uptake systems [154]. Organic acids like citrate and malate also seem to be involved in fungal iron uptake [155] as was shown by using a siderophore-free mutant of *Neurospora crassa* (*arg-5 ota aga*). This mutant grows well without added iron by secreting appreciable amounts of succinate, malate and citrate.

9 Siderophores of Pathogenic Fungi

9.1 Plant Pathogenic Fungi

In view of the physiological importance of iron it is of interest to see if there is any correlation between siderophore production and virulence in fungi. Plant pathogenic fungi are wide-spread in nature but there seems to be no indication that siderophore production prevails in the parasitic groups. *Ustilago sphaerogena* was the fungus from which the first siderophore, ferrichrome, was isolated and characterized [106]. Since then a variety of other smut fungi have been examined. Ferrichrome type siderophores seemed to be characteristic in the parasitic *Ustilaginaceae* and the closely related *Tilletiariaceae* while rhodotorulic acid prevailed in the saprophytic *Rhodosporidiaceae* [115]. It is interesting to note that the production of rhodotorulic acid from 42 isolates of smut fungi parasitizing members of the *Caryophyllaceae* led to the conclusion that *Ustilago violacea* could no longer remain in the genus Ustilago but was transferred to the separate genus *Microbotryum* [156]. Strains of *Fusarium* are also well known parasites of plants. The siderophores produced by *Fusarium* have been characterized and consisted mainly of fusarinins, dimerum acid and coprogen B [157, 122, 125]. However, no simple correlation could be found between siderophore production and pathogenicity when pathogenic and non-pathogenic strains were compared [158]. However, specific strains of root-colonizing bacteria, termed plant growth promoting rhizobacteria (PGPR), have recently been used as experimental inoculants to increase yield of plants [159]. The underlying theory is that pathogenic fungi and bacteria are suppressed by the presence of siderophores and other compounds excreted by the PRPG organisms resulting from a competition for iron between saprophytic and parasitic microorganisms.

9.2 Human Pathogenic Fungi

In animal systems there are also reports that fungal siderophores stimulate the growth of pathogens. *Escherichia coli* for example uses exogenous ferrichrome as an iron source [160, 161], although wild type strains develop multiple iron transport systems, e.g. the species-specific enterobactin transport system [162], the ferric citrate system [46] or the plasmid coded aerobactin system [29]. Moreover, receptors for the fungal coprogen and Fe-rhodotorulate have been identified in this genus [42]. These results support the view that enterobacterial pathogens profit remarkably from the use of exogenous siderophores. In most cases, however, the pathogen is able to biosynthesize its own siderophore in order to compete with the iron proteins of the host. Thus evidence has been presented that the pathogenic fungus *Histoplasma capsulatum* synthesized coprogen B during iron-limited conditions in a chemically defined medium [163] and it was shown that this siderophore exhibited growth factor activity for both the yeast and the mycelial form suggesting that siderophores might enhance fungal virulence in infected tissues by facilitating iron acquisition from iron-binding proteins, analogous to the mechanisms described for bacterial siderophores [164]. In a recent survey several other opportunistic and pathogenic fungi were examined for their capability to produce siderophores [165]. Hydroxamate type siderophores were detected in *Absidia corymbifera*, *Aspergillus niger*, *Rhizopus arrhizus*, *Rhizopus oryzae*, *Blastomyces dermatitidis*, *Histoplasma capsulatum*, *Sporothrix shenkii*, *Candida albicans* and *Trichophyton mentagrophytes*, although in this investigation only color formation was monitored. The dermatophyte *Microsporum gypseum* has been shown to produce ferricrocin as the principal siderophore under conditions of low-iron stress [166]. However, hydroxamate siderophores could not be found in *Saccharomyces cerevisiae* [167], suggesting that acidification of the medium provides a simple and yet very effective means of increasing the availability of iron and may be one of the devices used by *Ascomycetes* yeasts to achieve an adequate supply of the metal. The use of organic acids for iron acquisition in fungi [155] has been shown earlier by using a mutant strain of *Neurospora crassa* (*arg-5*, *ota*, *aga*), which is unable to synthesize siderophores in the absence of added ornithine [168]. In *Cryptococcus neoformans* two mechanisms for iron uptake were detected, one hydroxamate-independent and one hydroxamate-dependent, but the pathogen itself did not secrete hydroxamates [169]. Although it remains an open question whether these hydroxamates enter the cells as a whole, it is remarkable that siderophores of different origin can be used as an iron source in a fungus that obviously does not produce any siderophores.

10 Medical Applications of Siderophores

The previous chapters have demonstrated that the production of siderophores and their use for iron acquisition is of primary importance for the growth of bacteria and fungi in most environments, including man. The strong iron-withholding activity of human iron-binding proteins is counterbalanced by the siderophores which guarantee a sufficient iron supply under conditions of extreme deficiency of free iron. In the following, the potential and actual use of siderophores in therapy will be discussed.

10.1 Siderophores for Iron Chelation in Iron Overload Diseases

Iron overload diseases arise either by an increase in absorption from the gastrointestinal tract, as in primary idiopathic haemochromatosis, or by repeated blood transfusions to compensate congenital haemolytic anaemias. The thalassaemias seem to be the most frequent single-gene defects in the world population. They are a very diverse group of genetic disorders of haemoglobin synthesis. They are classified according to the type of globin chain whose synthesis is affected, that is as alpha, beta, beta/beta, and gamma thalassaemias.

Iron overload develops especially in patients suffering from thalassaemia major due to the enhanced absorption of dietary iron in an attempt to increase the low haemoglobin content. In humans there is no effective regulatory device to avoid iron accumulation by increased excretion. As a result an excess of 30 g of external iron will have accumulated at an age of 20 years. The mucosal hyperadsorption can be suppressed if the haemoglobin concentration is constantly maintained at 11.5 g/dl by transfusion. The hypertransfusion regimen used to compensate for ineffective erythropoesis leads to an accumulation of 3 to 4 g of iron per year. After 40 to 60 g of iron have been deposited in the reticuloendothelial and in the parenchymal cells, heart failure due to restricted cardiomyopathy, liver fibrosis and endocrine disorders become manifest.

The serious and frequently fatal consequences of iron overload can be prevented by high-dose chelation therapy with desferrioxamine B [170]. This is produced and marketed as the methanesulfonate salt by the Ciba-Geigy company under the brand name ®Desferal [8]. Desferrioxamine B is isolated from the culture of *Streptomyces pilosus*. It belongs to the siderophores of the hydroxamate type (Fig. 16). Desferrioxamine B has nearly ideal therapeutic properties. It exhibits a high affinity and selectivity for Fe^{3+}. Even after treatment for years with a dose of 5 g per week no toxicity or other serious side effects were encountered [171]. Desferrioxamine B is promptly excreted together with iron. Apparently, it does not cause redistribution of iron within the body nor does it withdraw iron form sites such as the respiratory chain where it fullfills an indispensable function.

Daily excretion of about 10 mg iron in urine and 20 mg in faeces was obtained using 0.5 mg desferrioxamine administered subcutaneously with a portable constant infusion pump 6 nights per week. This example shows that a negative iron balance can be achieved.

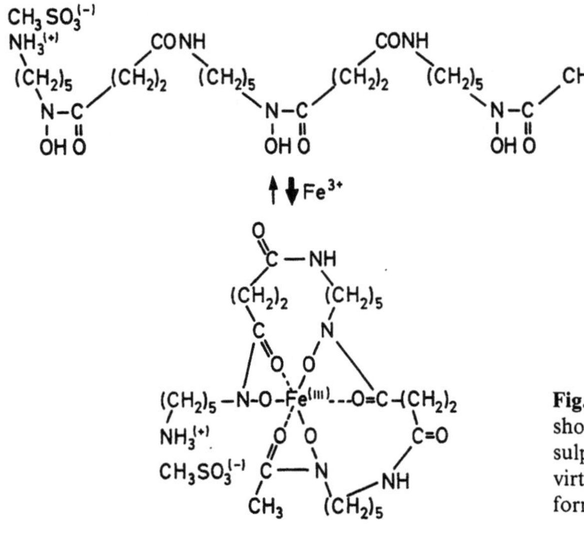

Fig. 16. Structure of desferrioxamine B shown as therapeutically used methanesulphonate salt. In the presence of Fe^{3+} virtually all iron exists in the complex form (strong arrow)

Both drug and the medical treatment are expensive. For this reason many siderophores of microbes and chemically synthesized iron chelating compounds have been tested to replace desferrioxamine B. Especially an orally — active drug was sought to avoid the cumbersome subcutaneous, intravenous, or intramuscular administration. Up to now, none was superior over desferrioxamine B. A new siderophore, designated desferrithiocin, was isolated from *Streptomyces antibioticus* [172]. In an animal model orally administered desferrithiocin substanially descreased hepatic iron overload in rats at concentrations which were nontoxic [173]. It remains to be seen whether this promising compound stands the clinical tests in comparison with desferrioxamine B. Synthetic substances such as enterochelin derivatives [51], pyridoxal isonicotinoylhydrazone, N,N'-bis(o-hydroxybenzyl)ethylenediamine diacetic acid, N,N'-ethylenebis(o-hydroxyphenylglycine) and hydroxypyridones are strong iron chelators and are being tested as orally active drugs [171].

10.2 Transport of Drugs Into Target Cells Via Fe^{3+}-Siderophore Carriers

In chapter B.II it was shown how the thionucleoside moiety of albomycin is taken up into cells of *E. coli* via the ferrichrome transport system. The iron chelating moiety serves as carrier for the thionucleoside (Fig. 1). The latter is cleaved from the iron carrier and is concentrated within the cells. Albomycin is an antibiotic for a wide range of gram-negative and gram-positive bacteria. Ferrichrome serves as an antagonist of albomycin and in the few cases studied in some detail, it competitively inhibits albomycin uptake. Albomycin is designated a sideromycin and ferrichrome is the corresponding siderophore, formely termed sideramine [170]. Another group of structurally related sideromycins and siderophores form the antibiotics A22765, ferrimycin A1 and desferrioxamine B.

Following the example of these compounds, antibiotics were chemically coupled to siderophores in order to bring them into target cells via active transport systems. In this way it was expected that the effective antibiotic dose could be lowered. Sulfanilamidonicotinic acid was coupled to ferricrocin (Fig. 9). The resulting ester inhibited *Staphylococcus aureus* cells at a concentration of 0.05 μM in contrast to the free sulfonamide for which 50 μM were required. Inhibition was antagonized by ferricrocin and by p-aminobenzoic acid suggesting uptake via the ferricrocin transport system and inhibition of the C1 metabolism by the sulfonamide derivative [174]. This example demonstrates a way by which potentially useful antibiotics, which enter cells too slowly by diffusion, could be converted to effective drugs. Uptake via active transport systems also lowers the effective dose which has to be administered so that potential side effects are minimized.

Very recently, a new cephalosporin, termed E-0702, was shown to be taken up by an iron transport system [175]. Iron-starvation which is known to derepress the iron transport systems, rendered cells of *E. coli* much more sensitive to the antibiotic. Moreover, *tonB* mutants were resistant to this cephalosporin. The side chain of the synthetic cephalosporin contains an ortho-dihydroxybenzene ring as is present in enterochelin (Fig. 4). It is likely but has not been shown that this antibiotic is taken up into *E. coli* via the enterochelin transport system. It was previously shown that synthetic enterochelin derivatives of rather different structures promote growth of *E. coli* on iron-limited media and enter cells via the enterochelin transport system [51]. MECAM is presented as an example in Fig. 4. Since the bactericidal activity of E-0702 against enteric bacteria was superior over other cephalosporins the concept of using iron transport systems to carry antibiotics into bacteria and fungi is worthy of further exploration.

10.3 Siderophores as Enzyme Inhibitors

Siderophores as potent iron chelators are able to inhibit iron-dependent enzymes by withrawing the metal. A number of studies have shown that the activity of the ribonucleotide reductase is decreased by microbial and synthetic siderophores [176]. As a consequence DNA synthesis is inhibited. Since strongly proliferating neoplastic cells frequently contain an increased density of receptors at the cell surface for the iron-delivering transferrin, they apparently have an enhanced iron requirement. Interference with the iron supply by administering siderophores may reduce their growth. In fact, this has been observed in vitro [6,176,177]. Therefore, siderophores are being investigated for their ability to control cell proliferation and their use as drugs in combination with other anticancer agents.

11 References

1. Archibald, F.: FEMS Microbiol. Lett. *19*, 29 (1983)
2. Hartmann, A. and Braun, V.: Arch. Microbiol. *130*, 353 (1981)
3. McIntosh, M. A. and Earhart, C. F.: J. Bacteriol. *131*, 331 (1977)
4. Spiro, T. G., Pape, L. and Saltman, P.: J. Amer. Chem. Soc. *89*, 5555 (1967)
5. Raymond, K. N., Müller, G. and Matzanke, B. F.: Topics in Current Chemistry *123*, 49 (1984)
6. Weinberg, E. D.: Physiol. Rev. *64*, 65 (1984)
7. Braun, V.: in: The Pathogenesis of Bacterial Infections (Jackson, G. G., Thomas, H., eds.) p. 168, Springer Verlag, Berlin 1985
8. Peter, H. H.: In: Hypertransfusion and Iron Chelation in Thalassaemia (Aksoy, M., Birdwood, G. F. B. eds.) p. 69, Huber Publishers, Bern 1985
9. Rosenberg, H. and Young, J. G.: In: Microbial Iron Metabolism (Neilands, J. A. ed.) p. 67, Academic Press, New York 1974
10. Wookey, P., Hussein, S. and Braun, V.: J. Bacteriol. *146*, 1158 (1981)
11. Benz, G., Schröder, T., Kurz, J., Wünsche, C., Karl, W., Steffens, G., Pfitzner, J., Schmidt, D.: Angew. Chem. *94*, 552 (1982)
12. Benz, G., Born, L., Brieden, M., Grosser, R., Kurz, J., Paulsen, H., Sinnwell, V., Weber, B.: Liebigs Ann. Chem. 1984, 1408
13. Braun, V., Günther, K., Hantke, K. and Zimmermann, L.: J. Bacteriol. *156*, 308 (1983)
14. Fecker, L. and Braun, V.: J. Bacteriol. *153*, 1301 (1983)
15. Coulton, J. W., Mason, P., Cameron, D. R., Carmel, G., Jean, R., Rode, H. N.: J. Bacteriol. *165*, 181 (1986)
16. Köster, W. and Braun, V.: Mol. Gen. Genet. 204, 435 (1986)
17. Burkhardt, R. and Braun, V.: Mol. Gen. Genet. *209*, 49 (1987)
18. Hoffmann, H., Fischer, E., Kraut, H. and Braun, V.: J. Bacteriol. *166*, 404 (1986)
19. Braun, V., Schaller, K. and Wolff, H.: Biochem. Biophys. Acta *323*, 87 (1973)
20. Hoffmann, H., Fischer, E., Schwartz, H. and Braun, V.: Arch. Microbiol. *145*, 334 (1986)
[illegible]. Ames, G. F. L.: Ann. Rev. Biochem. *55*, 397 (1986)
22. Bradbeer, C., Kenley, J. S., DiMasi, D. R. and Leighton, M.: J. Biol. Chem. *253*, 1347 (1978)
23. Heller, K. and Kadner, R. J.: J. Bacteriol. *161*, 104 (1985)

24. Friedrich, M. J., de Veaux, L. C. and Kadner, R. J.: J. Bacteriol. *167*, 928 (1986)
25. Gibson, F. and Magrath, D. J.: Biochim. Biophys. Acta *192*, 175 (1969)
26. Williams, P.: Infect. Immun. *26*, 925 (1979)
27. Williams, P. and Warner, P. J.: Infect. Immun. *29*, 411 (1980)
28. Stuart, S. J., Greenwood, K. T. and Luke, R. K. J.: J. Bacteriol. *143*, 35 (1980)
29. Braun, V.: FEMS Microbiol. Lett. *11*, 225 (1981)
30. Warner, P. J., Williams, P. H., Bindereif, A. and Neilands, J. A.: Infect. Immun. *33*, 540 (1981)
31. Gross, R., Engelbrecht, F. and Braun, V.: Molec. Gen. Genet. *201*, 204 (1985)
32. DeLorenzo, V. and Neilands, J. B.: J. Bacteriol. *167*, 350 (1986)
33. Carbonetti, N. H. and Williams, P. H.: Infect. Immun. *46*, 7 (1984)
34. Krone, W. J. A., Stegehuis, F., Konigstein, G., van Doorn, C., Roosendaal, B., de Graaf, F. K. and Oudega, B.: FEMS Microbiol. Lett. *26*, 153 (1985)
35. Braun, V., Gross, R., Köster, W. and Zimmermann, L.: Mol. Gen. Genet. *192*, 131 (1983)
36. Perez-Casal, J. F. and Crosa, J. H.: J. Bacteriol. *160*, 256 (1984)
37. McDougall, S. and Neilands, J. B.: J. Bacteriol. *159*, 300 (1985)
38. Bindereif, A. and Neilands, J. B.: J. Bacteriol. *161*, 727 (1985)
39. Colonna, B., Nicoletti, M., Visca, P., Cassalino, M., Valenti, P. and Maineone, F.: J. Bacteriol. *162*, 307 (1985)
40. Griffith, E., Stevenson, P., Hale, T. S. and Formal, S. B.: Infect. Immun. *49*, 67 (1985)
41. Lawlor, K. M. and Payne, S. M.: J. Bacteriol. *160*, 266 (1984)
42. Hantke, K.: Mol. Gen. Genet. *191*, 301 (1983)
43. Wagegg, W. and Braun, V.: J. Bacteriol. *145*, 156 (1981)
44. Frost, G. E. and Rosenberg, H.: Biochim. Biophys. Acta *330*, 90 (1973)
45. Zimmermann, L., Hantke, K. and Braun, V.: J. Bacteriol. *159*, 271 (1984)
46. Hussein, S., Hantke, K. and Braun, V.: Eur. J. Biochem. *117*, 431 (1981)
47. Hantke, K.: Mol. Gen. Genet. *182*, 288 (1981)
48. Hantke, K.: FEMS Microbiol. Lett. *15*, 83 (1982)
49. Lundrigan, M. D. and Kadner, R. J.: J. Biol. Chem. *261*, 10797 (1986)
50. Pierce, J. R. and Earhart, C. F.: J. Bacteriol. *166*, 930 (1986)
51. Heidinger, S., Braun, V., Pecoraro, V. L. and Raymond, K. N.: J. Bacteriol. *153*, 109 (1983)
52. Ecker, D. J., Matzanke, B. F. and Raymond, K. N.: J. Bacteriol. *167*, 666 (1986)
53. Coderre, P. E. and Earhart, C. F.: FEMS Microbiol. Lett. *25*, 111 (1984)
54. Fleming, T. P., Nahlik, M. D., Neilands, J. B. and McIntosh, M. A.: Gene *34*, 47 (1985)
55. Braun, V., Brazel-Faisst, C. and Schneider, R.: FEMS Microbiol. Lett. *21*, 99 (1984)
56. Moore, D. G., Yancey, R. J., Lankford, C. E. and Earhart, C. F.: Infect. Immun. *27*, 418 (1980)
57. Moore, D. G. and Earhart, C. F.: Infect. Immun. *31*, 631 (1981)
58. Fitzgerald, S. P. and Rogers, H. J.: Infect. Immun. *27*, 302 (1980)
59. Funakoshi, S., Doi, T., Nakajima, T., Suyama, T. and Tokuda, M.: Microbiol. Immunol. *26*, 227 (1982)
60. Konopka, K. and Neilands, J. B.: Biochemistry *23*, 2122 (1984)
61. Konopka, K., Bindereif, A. and Neilands, J. B.: Biochemistry *21*, 6503 (1982)
62. Martinez, J. L., Cercenado, E., Pérez-Diaz, J. C. and Baquero, F.: FEMS Microbiol. Lett. *37*, 259 (1986)
63. Achtman, M., Heuzenroeder, M., Kusecek, B., Ochman, H., Caugant, D., Selander, R. K., Vaisanen-Rhen, V., Korhonen, T. K., Stuart, S., Orskov, F. and Orskov, I.: Infect. Immun. *51* 268 (1986)
64. Carbonetti, N. H., Parry, S. H., Vaisanen-Rhen, V., Korkonen, T. K. and Williams, P.: Infect. Immun. *51*, 966 (1986)
65. Lam, C., Turnowsky, F., Schwarzinger, E. and Neruda, W.: FEMS Microbiol. Lett. *24*, 255 (1984)
66. Hacker, J. and Hughes, C.: Curr. Top. Microbiol. Immunol. *118*, 139 (1985)
67. West, S. E. H. and Sparling, P. F.: Infect. Immun. *47*, 388 (1985)
68. Mietzner, T. A., Barnes, R. C., Jeanlouis, Y., Shafer, W. M. and Morse, S. A.: Infect. Immun. *51*, 60 (1986)
69. DeVoe, I. W.: Microbiol. Rev. *46*, 162 (1982)
70. Mickelsen, P. A., Blackman, E. and Sparling, P. F.: Infect. Immun. *35*, 915 (1982)
71. Johnson, A. P.: FEMS Microbiol. Rev. *32*, 81 (1985)

72. Winkelmann, G.: In: Biotechnology, Vol. 4 (Rehm, H. J., Reed, G., eds.) p. 215, VCH Publ. Company, Weinheim, 1986
73. Moores, J. C., Magazin, M., Ditta, G. S. and Leong, J.: J. Bacteriol. *157*, 53 (1984)
74. Marugg, J. D., van Spanje, M., Hoekstra, W. P. M., Schippers, N. and Weisbeek, P. J.: J. Bacteriol. *164*, 563 (1985)
75. Hohnadel, D., Haas, D. and Meyer, J.-M.: FEMS Microbiol. Lett. *36*, 195 (1986)
76. Cox, C. D., Reinhart, K. L., Moore, M. L., Cook, J. C.: Proc. Natl. Acad. Sci. USA *78*, 4256 (1981)
77. Philson, S. B., Llinas, M. J.: J. Biol. Chem. *257*, 8086 (1982)
78. Teintze, M., Hossain, M. B., Barnes, C. L., Leong, J., van der Helm, D.: Biochemistry *20*, 6446 (1981)
79. Messenger, A. J. M., Hall, R. M. and Ratledge, C.: J. Gen. Microbiol. *132*, 845 (1986)
80. Hall, R. M. and Ratledge, C.: J. Gen. Microbiol. *133*, 193 (1987)
81. Snow, G. A.: Bacteriol. Rev. *34*, 99 (1970)
82. Braun, V.: In: The Enzymes of Biological Membranes (Martonosi, A. N. ed.) Vol. 3, p. 617, Plenum Press New York (1985)
83. Bennett, R. L., Rothfield, L. J.: J. Bacteriol. *127*, 498 (1976)
84. Payne, S. M., Niesel, D. W., Peixotto, S. S., Lawlor, K. W.: J. Bacteriol. *155*, 949 (1983)
85. Sigel, S. P., Payne, S. M.: J. Bacteriol. *150*, 148 (1982)
86. Carniel, E., Mazigh, D., Mollaret, H. H.: Infect. Immun. *55*, 277 (1987)
87. Shand, G. H., Anwar, H., Kadurugamuwa, J., Brown, M. R. W., Silverman, S. H., Melling, J.: Infect. Immun. *48*, 35 (1985)
88. Norquist, A., Davies, J., Norlander, L., Normark, S.: FEMS Microbiol. Lett. *4*, 71 (1978)
89. Sokol, P. A., Woods, D. E.: Infect. Immun. *40*, 665 (1983)
90. Meyer, J. M., Mock, M., Abdallah, M. A.: FEMS Microbiol. Lett. *5*, 395 (1979)
91. Chart, H., Trust, T. J.: J. Bacteriol. *156*, 578 (1983)
92. Crosa, J. H., Hodges, L. L.: Infect. Immun. *31*, 223 (1981)
93. Page, W. J., Huger, M.: J. Bacteriol. *158*, 496 (1984)
94. Wright, A., Simpson, L. M., Richardson, K., Maneval, D. R., Oliver, J. D., Morris, J. G.: FEMS Microbiol. Lett. *35*, 255 (1986)
95. Williams, P., Brown, M. R. W.: FEMS Microbiol. Lett. *33*, 153 (1986)
96. Matzanke, B. F., Bill, E., Müller, E., Trautwein, A. X. and Winkelmann, G.: Eur. J. Biochem. *162*, 643 (1987)
97. Emery, T.: In: Iron Transport in Microbes, Plants and Animals (Winkelmann, G., van der Helm, D. and Neilands, J. B., eds.) VCH Publ. Comp. Weinheim 1987
98. Neilands, J. B.: Ann. Rev. Biochem. *50*, 715 (1981)
99. Structure and Bonding 58, Springer, 1984
100. Winkelmann, G.: In: Iron, Siderophores, and Plant Diseases (Swinburne, T. R. ed.), Nato ASI Series, Plenum, New York London, 1986
101. Demange, P., Wendenbaum, S., Bateman, A., and Abdallah, M. A.: In: Iron Transport in Microbes Plants and Animals (Winkelmann, G., van der Helm, D., Neilands, J. B., eds.), VCH Publ. Comp., Weinheim 1987
102. Wong, G. B., Kappel, M. J., Raymond, K. N., Matzanke, B. and Winkelmann, G.: J. Am. Chem. Soc. *105*, 810 (1983)
103. Harris, W. R., Carrano, C. J., Raymond, K. N.: J. Am. Chem. Soc. *101*, 2213 (1979)
104. Emery, T.: Biochemistry *15*, 2733 (1976)
105. Powell, P. E., Cline, G. R., Reid, C. P. P. and Szaniszlo, P. J.: Nature *287*, 833 (1980)
106. Neilands, J. B.: J. Am. Chem. Soc. *74*, 4846 (1952)
107. Keller-Schierlein, W. and Maurer, B.: Helv. Chim. Acta *52*, 603 (1969)
108. Keller-Schierlein, W. and Deer, A.: Helv. Chim. Acta *46*, 1907 (1963)
109. Llinas, M. and Neilands, J. B.: Biophys. Struct. Mech. *2*, 105 (1976)
110. Emery, T. F. and Neilands, J. B.: J. Am. Chem. Soc. *83*, 1626 (1961)
111. Keller-Schierlein, W.: Helv. Chim. Acta *46*, 1920 (1963)
112. Jalal, M. A. F., Mocharla, R., Barnes, C. L., Hossain, M. B., Powell, D. R., Eng-Wilmot, D. L., Grayson, S. L., Benson, B. A. and van der Helm, D.: J. Bacteriol. *158*, 683 (1984)
113. Atkin, C. L. and Neilands, J. B.: Biochemistry *7*, 3734 (1968)
114. Widmer, J. and Keller-Schierlein, W.: Helv. Chim. Acta *57*, 1904 (1974)

115. Atkin, C. L., Neilands, J. B. and Phaff, H. J.: J. Bacteriol. *103*, 722 (1970)
116. Deml, G. and Oberwinkel, F.: Zentralbl. Bakteriol. Hyg. I. Abt. Orig. *C 3*, 475 (1982)
117. Carrano, C. and Raymond, K. N.: J. Am. Chem. Soc. *100*, 5371 (1978)
118. Scarrow, R. C., White, D. L. and Raymond, K. N.: J. Am. Chem. Soc. *107*, 6540 (1985)
119. Carrano, C. and Raymond, K. N.: J. Bacteriol. *136*, 69 (1979)
120. Müller, G., Barclay, S. J. and Raymond, K. N.: J. Biol. Chem. *260*, 13916 (1985)
121. Sayer, J. M. and Emery, T.: Biochemistry *7*, 184 (1968)
122. Diekmann, H.: Arch. Microbiol. *58*, 1 (1967)
123. Diekmann, H. and Krezdorn, E.: Arch. Microbiol. *106*, 191 (1975)
124. Adjimani, J. P. and Emery, T.: J. Bacteriol. *169*, 3664 (1987)
125. Diekmann, H.: Arch. Microbiol. *73*, 65 (1970)
126. Keller-Schierlein, W. and Diekmann, H.: Helv. Chim. Acta *53*, 2035 (1970)
127. Frederic, C. B., Bentley, M. D. and Shive, W.: Biochemistry *20*, 2436 (1981)
128. Frederic, C. B., Bentley, M. D. and Shive, W.: Biochem. Biophys. Res. Commun. *105*, 133 (1982)
129. Jalal, M. A. F., Mocharla, R. and van der Helm, D.: J. Chromatogr. *301*, 247 (1984)
130. van der Helm, D., Baker, J. R., Eng-Wilmot, D. L., Hossain, M. B. and Loghry, R. A.: J. Am. Chem. Soc. *102*, 4224 (1980)
131. Llinas, M., Klein, M. P. and Neilands, J. B.: J. Biol. Chem. *248*, 915 (1973)
132. Llinas, M., Wilson, D. M., Klein, M. P. and Neilands, J. B.: J. Mol. Biol. *104*, 853 (1976)
133. Emery, T.: Biochemistry *6*, 3858 (1967)
134. van der Helm, D., Jalal, M. A. F. and Hossain, M. B.: In: Iron Transport in Microbes, Plants and Animals (Winkelmann, G., van der Helm, D., Neilands, J. B., eds.) VCH Publ. Company, Weinheim 1987.
135. Naegeli, H. U. and Keller-Schierlein, W.: Helv. Chim. Acta *61*, 2088 (1978)
136. Huschka, H., Naegeli, H. U., Leuenberger-Ryf, H., Keller-Schierlein, W. and Winkelmann, G.: J. Bacteriol. *162*, 715 (1985)
137. Winkelmann, G.: FEBS Lett. *97*, 43 (1979)
138. Winkelmann, G. and Braun, V.: FEMS Microbiol. Lett. *11*, 237 (1981)
139. Winkelmann, G.: In: The Biological Chemistry of Iron (Dunford, H. B., Dolphin, D., Raymond, K. N., Sieker, L., eds.), D. Reidel Publ. Co., Dordrecht, The Netherlands 1981
140. Winkelmann, G.: Arch. Microbiol. *98*, 39 (1974)
141. Emery, T. and Emery, L.: Biochem. Biophys. Acta *50*, 670 (1973)
142. Ramasamy, K., Olsen, R. K. and Emery, T.: Abstr. 8th Amer. Peptide Sympos. 1983
143. Emery, T., Emery, L. and Olsen, R. K.: Biochem. Biophys. Res. Commun. *119*, 1191 (1984)
144. Huschka, H., Jalal, M. A. F., van der Helm, D. and Winkelmann, G.: J. Bacteriol. *167*, 1020 (1986)
145. Deml, G., Voges, K., Jung, G. and Winkelmann, G.: FEBS Lett. *173*, 53 (1984)
146. Jalal, M. A. F., Galles, J. L., van der Helm, D.: J. Org. Chem. *50*, 5642 (1985)
147. Chung, T. D. Y., Matzanke, B. F., Winkelmann, G. and Raymond, K. N.: J. Bacteriol. *165*, 283 (1986)
148. Horowitz, N. H., Charlang, G., Horn, G. and Williams, N. P.: J. Bacteriol. *127*, 135 (1976)
149. Ernst, J. F. and Winkelmann, G.: Biochim. Biophys. Acta *500*, 27 (1977)
150. Müller, G., Barclay, S. J. and Raymond, K. N.: J. Biol. Chem. *260*, 13916 (1985)
151. Ecker, D. J., Passavant, C. W. and Emery, T.: Biochim. Biophys. Acta *720*, 242 (1982)
152. Ecker, D. J., Lancaster, J. R., Emery, T.: J. Biol. Chem. *257*, 8623 (1982)
153. Matzanke, B. and Winkelmann, G.: FEBS Lett. *130*, 50 (1981)
154. Bienfait, H. F.: J. Bioenergetics Biomembr. *17*, 73 (1985)
155. Winkelmann, G.: Arch. Microbiol. *121*, 43 (1979)
156. Deml, G.: System. Appl. Microbiol. *6*, 23 (1985)
157. Emery, T.: Biochemistry *4*, 1410 (1965)
158. Lemanceau, P., Alabouvette, C. and Meyer, J. M.: In: Iron, Siderophores and Plant Diseases (Swinburne, T. R. ed.) Plenum Press, New York—London, 1986
159. Kloepper, J. W., Schroth, M. N. and Miller, T. D.: Phytopathology *70*, 1078 (1980)
160. Wayne, R. and Neilands, J. B.: J. Bacteriol. *212*, 497 (1975)
161. Hantke, K. and Braun, V.: FEBS Lett. *49*, 301 (1975)
162. McIntosh, M. A. and Earhart, C. F.: Biochem. Biophys. Res. Commun. *70*, 315 (1976)
163. Burt, W.: Infection and Immun. *35*, 990 (82)

164. Neilands, J. B., Bindereif, A. and Montgomerie, J. Z.: Current Topics in Microbiol. and Immun. *118*, 179 (1985)
165. Holzberg, M. and Artis, W. M.: Infect. Immun. *40*, 1134 (1983)
166. Bentley, M. D., Anderegg, R. J., Szaniszlo, P. J. and Davenport, R. F.: J. Am. Chem. Soc. *25*, 1455 (1986)
167. Neilands, J. B., Konopka, K., Schwyn, B., Coy, M., Francis, R. T., Paw, B. H. and Bagg, A.: In: Iron Transport in Microbes, Plants and Animals (Winkelmann, G. van der Helm, D., Neilands, J. B. eds.) VCH Publ. Co., Weinheim 1987
168. Winkelmann, G. and Zähner, H.: Arch. Microbiol. *88*, 49 (1973)
169. Jacobson, E. S. and Petro, M. J.: J. Med. and Vet. Mycology (in press)
170. Bickel, H., Bosshard, R., Gäumann, E., Reusser, E., Vischer, E., Vosser, W., Wettstein, A. and Zähner, H.: Helv. Chim. Acta. *43*, 2119 (1960)
171. Hypertransfusion and iron chelation in thalassaemia (Aksoy, M., Birdwood, G. F. B., eds.) Hans Huber Publishers, Bern, Stuttgart, Toronto 1985
172. Naegeli, H. U. and Zähner, H.: Helv. Chim. Acta *63*, 1400 (1980)
173. Longeville, A. and Crichton, R. R.: Biochem. Pharmacol. *35*, 3669 (1987)
174. Zähner, H., Diddens, H., Keller-Schierlein, W. and Naegeli, H. U.: Jap. J. Antibiot. 30, Suppl. 201–206 (1977)
175. Watanabe, N. A., Nagasu, T., Katsu, K. and Kitoh, K.: Antimir. Ag. Chemother. *31*, 497 (1987)
176. Bergeron, R. J.: Trends in Biochemical Sciences *11*, 133 (1986)
177. Kontoghiorges, G. J., Piga, A., Hoffbrand, A. V.: FEBS Lett. *204*, 208 (1986)

Pharmacokinetic Drug Interactions

W. R. Bartle, Pharm. D., S. E. Walker, M. Sc. Phm., N. E. Winslade, Pharm. D.

Department of Pharmacy, Sunnybrook Medical Centre and The Faculty of Pharmacy, University of Toronto, Toronto, Ontario, Canada

The clinical observation that one drug may affect the disposition of a co-administered drug has led to extensive research of pharmacokinetic drug interactions. Drug absorption, distribution, metabolism and/or excretion may be modified by the concurrent administration of other therapeutic agents. This review discusses the various pharmacokinetic mechanisms involved in these changes and discusses several drug and patient-specific factors that aid in assessing the clinical significance of these interactions.

Progress in Clinical Biochemistry
and Medicine, Vol. 5

1 Introduction

A drug-drug interaction occurs when two drugs given together affect the pharmacokinetics and/or pharmacodynamics of each other. The potential for drug interactions to occur is practically limitless. Robinson [1] estimated that an average hospital drug formulary consisting of 200 items has theoretically 19,900 different paired drug combinations. Many patients, however, receive more than two drugs at any one time [2,3] and thus, the likelihood of an interaction may increase beyond an estimate of drug pairings.

A study of computerized prescription histories of nearly 42,000 patients in California suggested that drug interactions could have occurred in 7.6 percent of the cases. Moreover, greater than 50 percent of patients receiving oral anticoagulants were taking additional drugs known to interact with them [4]. Stewart and Cluff [5] interviewed 75 patients taking prescription and non-prescription drugs and found that 51.7 percent of the patients had taken a combination known to interact. Neither study, unfortunately, determined how many patients actually developed a clinically significant problem as a result of multiple drug use. Despite this shortcoming, these epidemiologic studies serve to emphasize the magnitude and potential health consequences of drug interactions.

Drugs may interact on a pharmaceutical, pharmacodynamic or pharmacokinetic basis. Pharmaceutical interactions occur when one drug is chemically inactivated by the addition of another. A well-known example of such a pharmaceutical interaction is the inactivation of an aminoglycoside antibiotic by the addition of a penicillin to the intravenous fluid [6].

Pharmacodynamic interactions arise when drugs act upon similar or competing receptors or physiologic systems. A recent example of this type of interaction is the observed decrease in the effect of several anti-hypertensive drugs by concomitant administration of a non-steroidal anti-inflammatory drug [7].

Pharmacokinetic interactions occur when one drug modifies the absorption, distribution, metabolism and/or elimination of another drug. Generally speaking, a drug affects only one pharmacokinetic parameter of the other drug, although exceptions to this have been documented [8]. In addition, a single drug may affect the pharmacokinetics of several different drugs by various means (Table 1). Interactions are usually reported as unidirectional in that only one of the drugs is apparently affected and one is not. However, this may be due, in part, to the fact that most published studies have only examined the disposition of one of the interacting drugs. It has been shown, though, that iron supplements and tetracycline, affect the absorption of each [20], and that the simultaneous administration of cimetidine and, cisapride, a drug that increases gastrointestinal motility, results in the altered pharmacokinetics of both drugs [21].

Pharmacokinetic drug interactions are frequently therapeutically detrimental as they may either alter drug toxicity, alter pharmacologic response or reduce efficacy. Occasionally, these interactions are intentionally advantageous. For example, probenecid, an uricosuric agent, decreases the active renal secretion of ampicillin and enhances the latters effectiveness when used as a single dose regimen for the treatment of gonorrhea.

Table 1. Examples of multiple mechanisms of pharmacokinetic interactions by a single drug

Interacting drug	(Re)Absorption		Protein binding		Hepatic metabolism		Renal excretion	
	Increase	Decrease	Increase	Decrease	Increase	Decrease	Increase	Decrease
Antacids		digoxin [9]					salicylate [10]	
Cimetidine		ketoconazole [11]				phenytoin [12]		procainamide [13]
Phenytoin				thyroxine [14]	carbamazepine [15]			
Phenylbutzone				warfarin [16]		warfarin [17]		
Erythromycin	digoxin [18]					warfarin [19]		

Numbers refer to References

The remainder of this article focuses on the mechanisms and clinical relevance of primarily adverse pharmacokinetic drug interactions. Some examples of the various types of pharmacokinetic drug interactions will be presented throughout the article; however, the reader is directed to two recent references for a comprehensive list and evaluation of the numerous published drug interactions [22,23].

1.1 Classification of Pharmacokinetic Drug Interactions

Pharmacokinetic drug interactions may be classified according to several different categories, including site of interaction, mechanism of interaction or significance of interaction. From a physiologic viewpoint, a classification based on the site of interaction, as depicted in Fig. 1, is most logical. This classification, however, provides no insight into the mechanism of, or the clinical significance of the interaction. A classification categorized into absorption, distribution, metabolism and elimination (ADME), is not only familiar to most clinicians, but also provides a background for

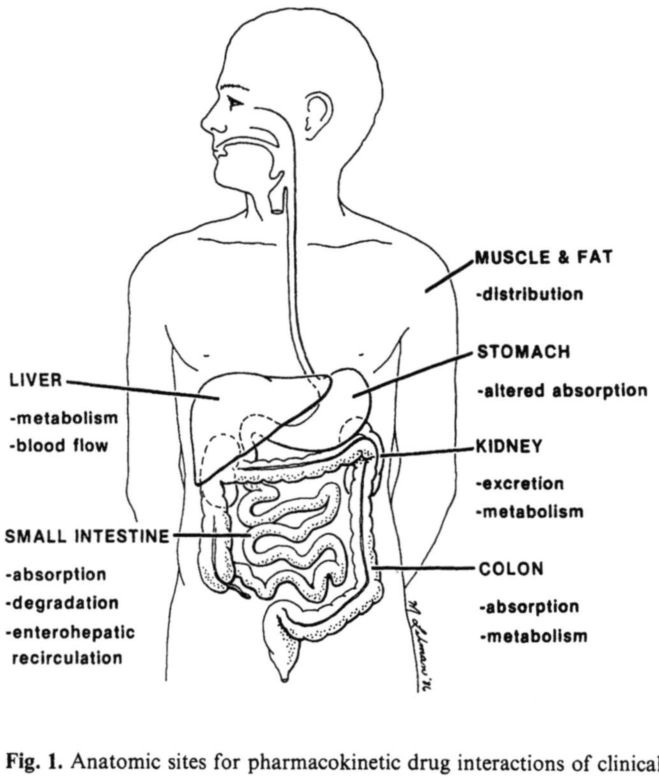

Fig. 1. Anatomic sites for pharmacokinetic drug interactions of clinical significance

the understanding of the mechanisms involved in these interactions. Once the mechanism is understood, both the site and the clinical significance may be determined. In order to maximize the understanding of pharmacokinetic drug interactions, the remainder of this review will be classified according to the ADME scheme, as outlined in Table 2.

Table 2. Classification of pharmacokinetic drug interactions

1. Altered Absorption	4. Altered Elimination
a) Drug-induced changes in gastrointestinal motility	a) renal excretion
b) Drug-interactions directly affecting absorption	i) filtration
2. Altered Metabolism	ii) reabsorption
a) hepatic enzyme activity	iii) secretion
b) hepatic blood flow	b) renal metabolism
3. Altered Distribution	
a) plasma protein binding	
b) tissue protein binding	

2 Altered Gastrointestinal Absorption

Oral preparations account for the majority of all drugs administered. Thus, the gastrointestinal tact and its contents play a major role in determining the rate and extent of drug absorption. The effect of food on drug absorption and other pharmacokinetic parameters will not be dealt with in this article, but has recently been reviewed by Melander [24].

Major factors that determine intestinal drug absorption are the physiochemical characteristics of the drug and its dosage form and the physiologic characteristics at or near the site of drug absorption. A drug must be in solution before absorption can take place, and generally speaking, tablet and capsule formulations must disintegrate before the drug they contain can undergo dissolution. With some new osmotically-controlled oral formulations, true disintegration is not required [25]. Under normal conditions, the extent of drug absorption in the stomach is unknown but is probably modest. The small intestine is the most important site for drug absorption in the gastrointestinal tract because the mucosal surface area through which absorption can take place is extraordinarily large. Due to of the presence of villi and microvilli, the surface area in man has been estimated to range from 80 cm^2/cm serosal length just beyond the duodenaljejunal flexure to about 20 cm^2/cm serosal length just before the ileocecal valve [26]. Drugs also undergo limited absorption from the colon [27], but this is of minor importance for most oral formulations. According

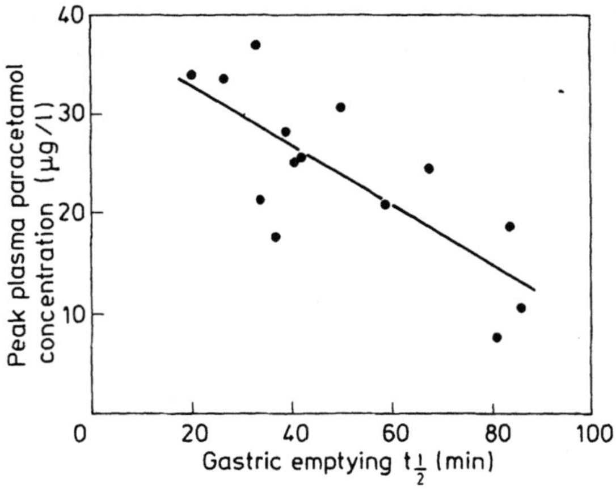

Fig. 2. Relationship between the plasma concentration and gastric emptying half-time in different patients after oral administration of 1.5 g acetaminophen (from Ref. [28] with permission)

to the pH partition hypothesis, only nonionized, lipophilic drugs penetrate the intestinal mucosa. Therefore, gastrointestinal pH could influence the rate and extent of absorption of ionizable drugs across the epithelium.

The rate of drug absorption is controlled primarily by gastric emptying time (Fig. 2) which influences two pharmacokinetic parameters, T_{max}, the time to reach peak plasma concentration, and to some extent, C_{max}, the peak plasma concentration [28]. Welling [29] has classified drug-drug absorption interactions in terms of indirect effects of one drug on gastrointestinal tract physiology or direct interactions involving altered pH, adsorption, absorption or chelation. Most, but not all of these drug-drug interactions result in reduced or delayed systemic absorption.

2.1 Drug-Induced Changes in Gastrointestinal Motility and Function

Anticholinergic drugs, propantheline being the most studied example, decrease the gastric emptying rate and have been shown to increase the extent of absorption of hydrochlorothiazide [30] and digoxin [31] (Fig. 3). Metoclopramide, a drug which increases gastric emptying rate, has the opposite effect on the absorption of these drugs. These observations, plus the lack of an effect on the extent of absorption of digoxin administered as a liquid formulation, suggest that the time allowed for tablet dissolution was increased by the anticholinergic agent and decreased by metoclopramide. Aluminium ion, a common component of antacids, is known to delay both gastric emptying [32] and the absorption of many drugs [29]. Neomycin can increase gastric emptying time and we have shown it can delay the absorption of spironolactone [33], a drug frequently prescribed with this antibiotic (Fig. 4).

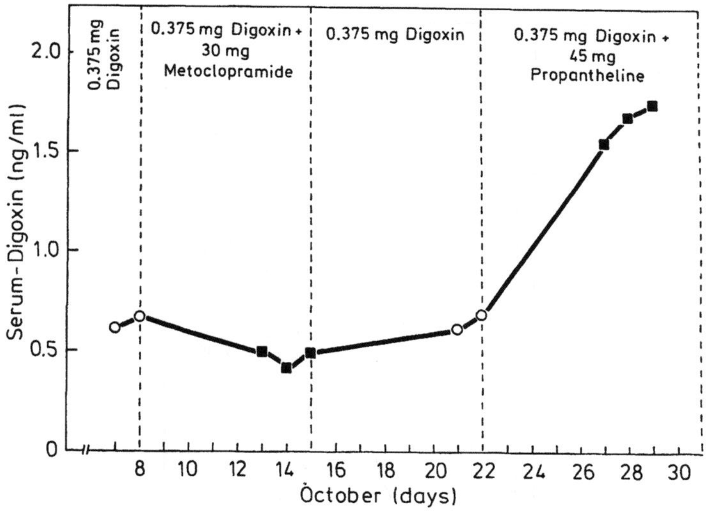

Fig. 3. Variations in serum-digoxin concentration in a patient during treatment with digoxin and metoclopramide or propantheline (from Ref. [31]) with permission)

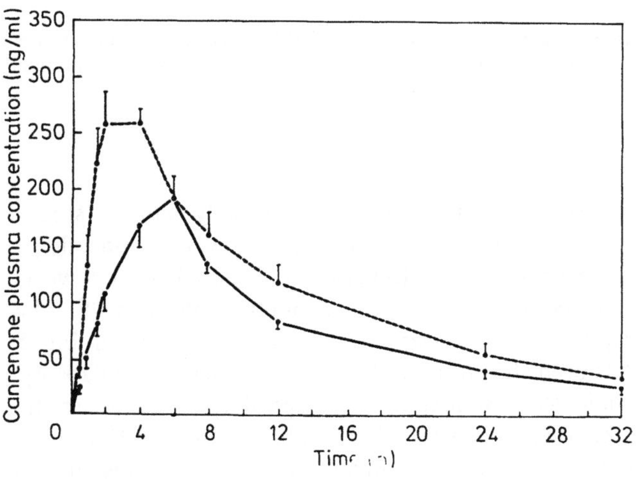

Fig. 4. Mean plasma canrenone concentrations versus time curves for 11 males following spironolactone (– – –) and spironolactone plus neomycin (———). Peak plasma canrenone concentration was significantly longer after 1 g neomycin. Canrenone is a major circulating metabolite of spironolactone (from Ref. [33]) with permission)

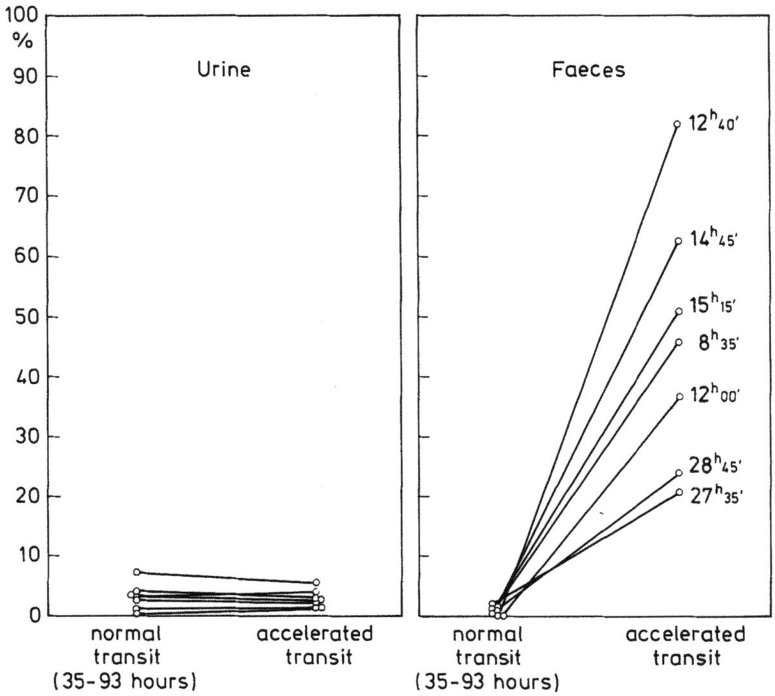

Fig. 5. Urinary and faecal recoveries of salicylazosulphapyridine (SASP) in healthy volunteers after ingestion of a single dose of SASP during a normal intestinal transit and after an accelerated intestinal transit induced by bisacodyl (from Ref. [34] with permission)

Since the contribution of the colon to drug absorption is usually minimal, studies examining drug interactions in the lower intestine are few; however, bisacodyl, a colono-active laxative [34], has been shown to decrease the bacterial breakdown of sulphasalazine to its components, sulphapyridine and 5-aminosalicylate [34]. This results in the decreased absorption of the sulphapyridine moiety as depicted in Fig. 5.

Chronic oral neomycin has been shown to produce a gastrointestinal malabsorption syndrome which is pathologically similar to non-tropical sprue. However, the mechanism underlying the decrease in the extent of absorption of digoxin [35] remains unknown as that study involved only a single dose of neomycin, a dose that is unlikely to produce a histologic change in the mucosa.

2.2 Drug Interactions Directly Affecting Absorption

Antacids may alter the absorption of drugs or their dosage forms in various ways. By raising gastric pH they affect the dissolution rate of some solid dosage forms or the degree of ionization of drugs that are weak acids or bases. Antacids may also chelate or adsorb drug molecules or affect gastric emptying.

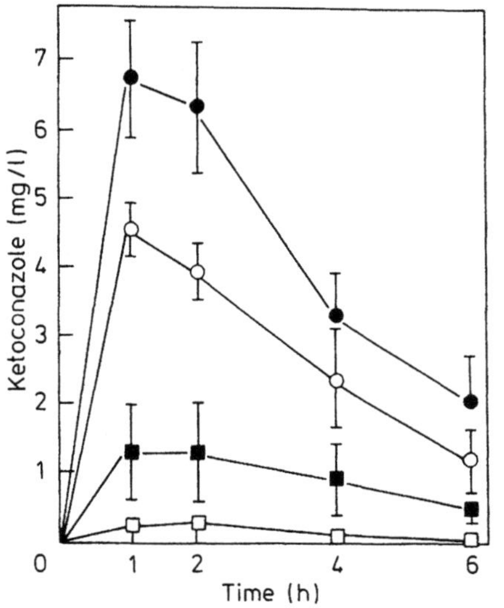

Fig. 6. Antifungal plasma levels in three volunteers receiving ketoconazole. ○, Fasting, 200 mg ketoconazole; ■, Fasting, 400 mg cimetidine. 200 mg ketoconazole. 2 h later; ●, Fasting, 400 mg cimetidine. 200 mg ketoconazole in acidified solution 2 h later. □, Fasting, 400 mg cimetidine. 100 mL 0.5 $NaHCO_3$, and 200 mg ketoconazole 2 h later (only determined in 2 volunteers). Bars indicate s.d. (from Ref. [37] with permission)

As reviewed by Welling [29], antacids have been shown to reduce the absorption of chlorpromazine, digoxin, isoniazid, penicillamine, and vitamin A. In most cases, reduced drug availability is attributed to adsorption properties of the antacids. Sodium bicarbonate, however, decreased the absorption of tetracycline from a capsule but not a liquid formulation suggesting the dissolution rate was affected [36]. Similarly, cimetidine has been shown to decrease the absorption of ketoconazole [37] by lowering the amount of gastric acid present, which is required for adequate solubilization of this dibasic lipophilic antifungal agent (Fig. 6). The rate of absorption of certain drugs from enteric-coated formulations is increased by concurrent administration of antacids [38], presumably by causing earlier and undesired dissolution of the pH-dependent coating.

The kaolin and pectin components of a commonly used anti-diarrheal preparation can physically adsorb many drugs and, as such, has led to reports of decreased absorption of lincomycin, tetracycline and digoxin [29].

The ability of activated charcoal to adsorb many different drugs and thus decrease their adsorption [39] has resulted in its increased use in the early management of drug overdose cases. A similar effect on drug re-absorption by charcoal appears to occur in the lower gastrointestinal tract for systemic drug that reaches the lumen by biliary secretion or other means [40]. The bile-acid binding resins, cholestyramine and colestipol, appear to possess similar adsorptive properties as activated charcoal, decreasing drug adsorption and re-absorption [41-43].

The interactions between certain therapeutically used metal ions and oral tetracyclines is one of the earliest and best documented pharmacokinetic drug interactions affecting absorption and has been reviewed extensively by Neuvonen [44]. The basis for the interaction is the formation of drug-metal complexes that are either insoluble and/or poorly diffusable. The stability of the tetracyline/metal complexes are, in order of decreasing stability and increasing 'adsorbability': Fe^{+++}, Al^{+++}, Fe^{++}. The affinity of various tetracycline derivatives to cations can also be quite variable. For example, the binding of doxycyline to calcium in vitro is considerably less than many of the other tetracyclines. However, the absorption of all tetracycline derivatives, including doxycycline and minocycline, can be significantly decreased by the simultaneous intake of aluminium-magnesium antacids and iron supplements. A time interval of three hours between the administration of the tetracycline and the metal ion can be used to minimize this interaction.

2.3 Other Absorption Interactions

Lindenbaum et al. [45] have found that, in a small proportion of subjects, fecal bacteria may convert digoxin to reduced, cardioinactive metabolites. Three such subjects were given either concurrent tetracycline or erythromycin for five days. The amount of digoxin reduced metabolites in the urine was observed to decrease significantly and the steady-state digoxin plasma level increased, approximately doubling in two of the three subjects. However, these same investigators have shown that neomycin inhibits the absorption of digoxin. No generalization can thus be made concerning the affect of antibiotics on the absorption of this cardiac glycoside.

3 Altered Hepatic Metabolism

In order to understand the pharmacokinetic interactions which result from either altered enzyme activity or hepatic blood flow, it is necessary to review several relevant pharmacokinetic concepts. Clearance refers to the portion of the distribution volume from which the drug is completely removed per unit time. For example, if the clearance of a drug is 60 mL/min, this implies that the drug is completely removed from 60 mLs of "plasma" every minute by the eliminating organ(s). Although clearance may occur via several different routes, it is often divided into two nonspecific pathways; renal and nonrenal clearance. The former includes filtration and/or secretion, while the latter includes both metabolism and nonrenal excretion, such as volatile exhalation, or secretion into sweat or saliva. Although nonhepatic metabolism is important for some drugs, such as nitroglycerin (metabolized by vascular endothelium [46]) and prostaglandin E1 (hydrolyzed by lung esterases [47]), the majority of drug metabolism occurs in the liver. For ease of explanation, the principles discussed in the remainder of this section are based on the assumption that the parent drug is exclusively metabolized by the liver, such that total body clearance equals hepatic clearance. This assumption holds true for many drugs such as antipyrine, propranolol, lidocaine and tricyclic antidepressants [48]).

Total body clearance (TBC) may be calculated by several different manners, the most well known being designated by the equation:

$$\mathrm{TBC} = (\mathrm{K})\,(\mathrm{Vd}) \tag{1}$$

where Vd is the volume of distribution and K the elimination rate constant. Since TBC and VD are independent parameters, they may both affect the elimination rate constant and thus half-life of a drug. Clearance may also be calculated using equations 2 or 3:

$$TBC = \frac{dose_{iv}}{AUC_{iv\,0-\infty}} \tag{2}$$

$$TBC = \frac{F \times dose_{po}}{AUC_{po\,0-\infty}} \tag{3}$$

where AUC is the area under the intravenous (iv) or oral (po) drug plasma concentration versus time profile, and F is the bioavailability of the oral product. Bioavailability involves both the fraction of drug absorbed and the fraction of drug escaping the first pass effect of the liver. This latter concept of first pass relates to the efficacy of hepatic extraction, expressed as the extraction ratio (E). If one assumes that a drug is completely absorbed, then the fraction absorbed equals one, and bioavailability depends solely on the hepatic extraction ratio, such that:

$$F = 1 - E \tag{4}$$

Finally, for drugs which are exclusively metabolized, hepatic clearance (Clh) may be estimated using:

$$Clh = \frac{Qh \times Clint}{Qh + Clint} \tag{5}$$

where Qh is hepatic blood flow and Clint is the intrinsic clearance of the drug.

Intrinsic clearance is a term first defined by Branch et al. referring to the maximal ability of the liver to irreversibly remove drug by all pathways, in the absence of any hepatic blood flow restrictions [49]. Intrinsic clearance therefore reflects the inherent ability of the liver to remove drug and involves hepatocellular function, enzyme quantity, substrate-enzyme affinity and plasma protein binding. Examining this definition would suggest that intrinsic clearance should be related to the extraction ratio, since both involve the efficiency of drug removal by the liver. Equation 6 expresses the mathematical relationship between Clint and E:

$$E = \frac{Clint}{Qh + Clint} \tag{6}$$

Based on Equation 5, it is apparent that hepatic clearance may be altered by drug or disease-induced changes in either intrinsic clearance (via enzyme induction/inhibition or alterations in protein binding), or hepatic blood flow. The overall effect depends on the physiochemical properties of the individual drug being examined, the route of administration, and the degree of alteration of intrinsic clearance and/or hepatic blood flow.

3.1 Hepatic Enzyme Activity

There are a variety of drugs used that may either induce or inhibit hepatic enzyme activity (Table 3). Whether these agents will cause a significant pharmacokinetic interaction through altered enzyme activity depends mainly on the intrinsic clearance of the second, co-administered drug.

If this second drug has a low intrinsic clearance, its hepatic clearance will be predominantly affected by changes in enzyme activity and not hepatic blood flow. Low intrinsic clearance implies that only a small fraction of the total amount of drug being delivered to the liver per minute is removed. The liver therefore, has a capacity-limited ability to "clear" that drug from the plasma. The rate limiting step for hepatic clearance is therefore not the rate of delivery of drug to the liver (Qh), but the liver's ability to clear drug from the plasma. It therefore follows that increasing or decreasing hepatic blood flow, thereby changing the amount of drug presented to the liver, should not alter the clearance of this drug. Otherwise stated, the hepatic clearance of a low intrinsic clearance drug is flow-independent. What the hepatic clearance is dependent on is the efficiency of drug removal. Thus, altering the quantity of function of enzymes will alter hepatic clearance.

Mathematically, this concept can be demonstrated as follows:

(a) If Clint is low, then it may be assumed that Clint $\lll$ Qh, therefore

(b) If $\text{Clh} = \dfrac{\text{Qh} \times \text{Clint}}{\text{Qh} + \text{Clint}}$ (5)

and Clint is negligible with respect to Qh, then Eq. (5) reduces to

(c) $\text{Clh} = \text{Clint}$ (7)

According to Eq. (7), the hepatic clearance of low intrinsic clearance drugs should therefore be affected by factors such as enzyme or protein binding alterations, and not by drugs or conditions which alter hepatic blood flow.

This concept has been described by Wilkinson and Shand [59]. Figure 7, the left hand side, illustrates the effects of enzyme induction on the concentration time profile of a low intrinsic clearance drug. Since the clearance of this low intrinsic clearance drug is directly affected by enzyme induction, and assuming the volume of distribution remains constant, an increase in clearance is reflected by a shorter elimination half-life. According to Eq. (2) and reflected in Fig. 7, the area under the curve should vary inversely with the change in clearance.

Table 3. Examples of drugs that inhibit or induce hepatic drug metabolism

Inhibit	Induce
cimetidine [50]	phenobarbital [55]
chloramphenicol [51]	rifampicin [56]
disulfiram [52]	carbamazepine [57]
phenylbutazone [53]	spironolactone [58]
vaccines [54]	phenytoin [15]
erythromycin [19]	

Numbers refer to references

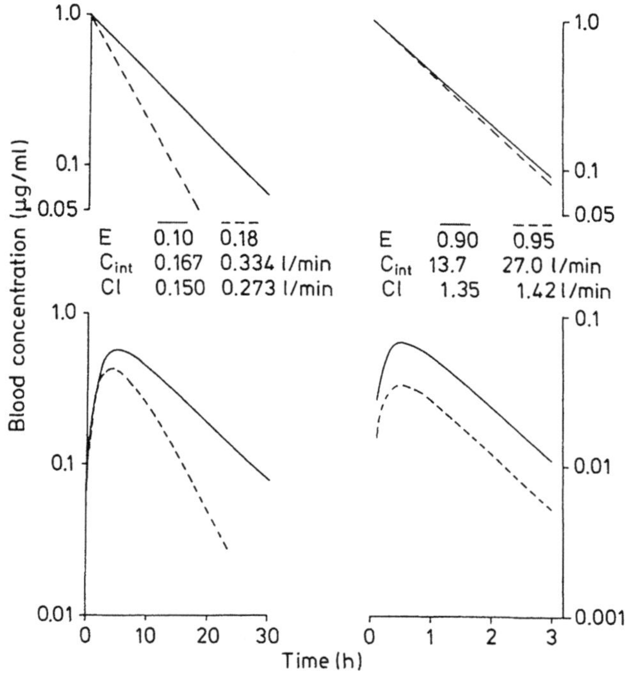

Fig. 7. The effect of increasing hepatic total intrinsic clearance ($Cl_{intrinsic}$) on the total blood concentration time curve after intravenous (upper panels) and oral (lower panels) administration of equal doses of two totally metabolized drugs. The left panels refer to a drug with an initial $Cl_{intrinsic}$ equivalent to an extraction ratio of 0.1 at a liver blood flow of 1.5 L/min and the right panels to one with an initial extraction ratio of 0.9 (from Ref. [59]) with permission)

Following oral administration (left lower graph of Fig. 7) the clearance and half-life are once again altered by enzyme induction. Unlike parenteral drug administration, however, the area under the curve after oral dosing is susceptible to enzyme-induced changes in systemic bioavailability. Enzyme induction will increase the efficiency of extraction on first pass with a resulting small decrease in bioavailability. Rearrangement of Eq. (3) to:

$$AUC_{po} = \frac{F \times dose_{po}}{TBC} \tag{8}$$

illustrates that this faster clearance and slightly decreased bioavailability would result in a significantly smaller area under the curve; given that:

$$CPss = \frac{F \times dose_{po}}{AUC_{po \times \psi}} \tag{9}$$

where ψ is the dosing interval, the average concentration at steady-state (CPss) will also decrease.

Several examples of these types of interactions exist, including the increased clearance and subsequent decreased effect of warfarin when either of the enzyme-

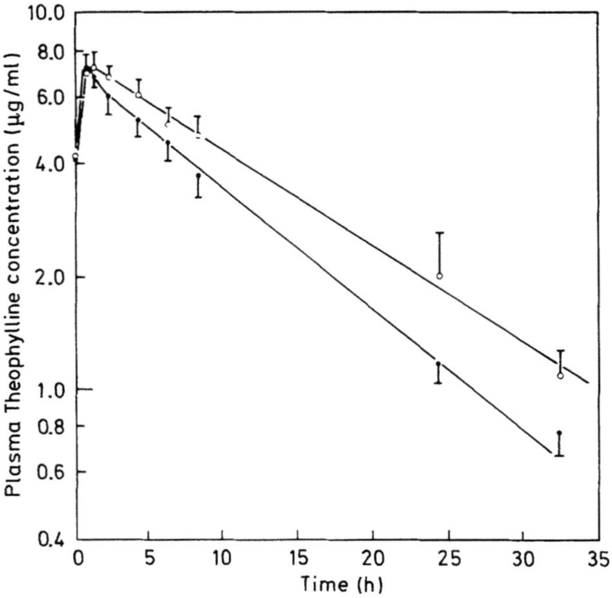

Fig. 8. Mean plasma theophylline concentration-time curves for concentrations obtained during administration of theophylline alone (●) and theophylline plus cimetidine (○). All points represents the mean ± standard error of 10 determinations. Originally published in Clin. Pharm., Vol. 2, Copyright (©) 1983, American Society of Hospital Pharmacists, Inc. All rights reserved. Reprinted with permission, Ref. [65]

inducers phenobarb [60] or rifampicin [61] are administered concomitantly. Conversely, a decrease in theophylline clearance is observed with the addition of an enzyme inhibitor such as erythromycin [62,63] or cimetidine [64,65]. Fig. 8 illustrates the interaction between low dose (400 mg qhs) cimetidine and oral theophylline in 10 healthy volunteers [65]. The changes in Fig. 8 reflect those predicted in the left hand lower graph of Fig. 7. Since theophylline is a "low intrinsic clearance" drug, enzyme inhibition by cimetidine causes a decrease in clearance (55.44 to 48.69 mL/min), an increase in area under the curve (101.30 to 118.88 µg. hr/mL respectively) and a prolongation of half-life (9.42 to 11.89 hr respectively).

For high intrinsic clearance drugs such as lidocaine and propranolol, the effect of enzyme induction or inhibition is very different. High intrinsic clearance implies that the drug is cleared by the liver very efficiently and therefore a high fraction of the drug presented to the liver is rapidly removed. In otherwords, the liver clears the drug as fast as it is delivered by the hepatic blood supply. The limiting factor to hepatic clearance therefore is the rate at which drug is presented to the liver. Clearance, therefore, would be affected by changes in liver blood flow and not intrinsic clearance changes.

Again, mathematically, this can be shown by:

(a) if Clint is high, then Qh $\lll$ Clint, and since

(b) $$\text{Clh} = \frac{\text{Qh} \times \text{Clint}}{\text{Qh} + \text{Clint}} \quad (5)$$

if Qh is negligible with respect to Clint, then equation 5 reduces to

(c) $$\text{Clh} = \text{Qh} \quad (10)$$

The right hand side of Fig. 7 illustrates the effects of enzyme induction on the intravenous (upper) and oral (lower) plasma concentration time profiles of a high intrinsic clearance drug. For intravenous administration, the clearance and half-life are unaffected by changes in intrinsic clearance. Following oral administration, the clearance and half-life again are not affected (as is shown by the parallel slopes of the terminal phases), but the area under the curve and steady state concentration are decreased by enzyme induction. This is explained by consideration of changes in the efficacy of hepatic extraction. Examining equations (4) and (6), the efficiency of extraction of a high intrinsic clearance drug is very high, such that the extraction ratio approaches one, and bioavailability is small. Although increasing intrinsic clearance causes only a slight increase in extraction ratio and a resulting small decrease in the absolute value of bioavailability, this results in a significant decrease in percent bioavailability. According to Fig. 7, the bioavailability of the sample drug decreased by only 0.05 percent (0.10 to 0.05%), yet this represents an overall 50 percent decrease in systemic bioavailability. This significantly lower bioavailability, in the face of unchanged clearance and volume of distribution, results in an identically shaped concentration-time curve, but a smaller area under the oral concentration time curve and a reduced steady-state concentration.

Examples of the differing effects of altered intrinsic clearance on intravenously versus orally administered high intrinsic clearance drugs are infrequent. Herman et al. [66] examined the enzyme inducing effect of rifampicin on the clearance of orally administered propranolol, a high intrinsic clearance drug. Propranolol clearance was determined at steady-state prior to and after three weeks of rifampicin treatment. Rifampicin caused significant decreases in propranolol concentrations at steady-state, without significant effects on elimination half-life. These effects parallel those illustrated in Fig. 7, the lower right hand graph. In contrast, the clearance of intravenous propranolol has been shown to be unaffected by hepatic enzyme inhibition secondary to chlorpromazine administration [67]. Bioavailabilty and thus plasma concentrations of orally administered propranolol, however, increased with chlorpromazine pretreatment. Neither half-life, volume of distribution nor protein binding were affected by this interaction, again reflecting the principles detailed in the lower right hand graph of Fig. 7.

3.2 Hepatic Blood Flow

Since there are a limited number of drugs which affect hepatic blood flow, the number of documented interactions based on this mechanism are few. As with altered intrinsic clearance however, the possibility of an interaction between an agent which decreases hepatic blood flow and a second drug depends on the intrinsic clearance and route of administration of this latter drug.

As previously discussed for a low intrinsic clearance drug, the rate limiting step for hepatic clearance is not the rate of drug delivery to the liver, but the inherent

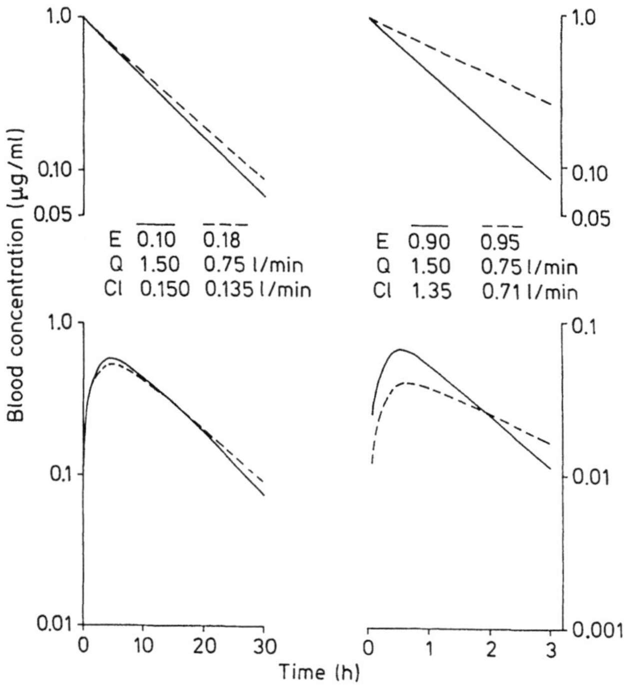

Fig. 9. Effects of decreasing liver blood flow on the total blood concentration/time curves after intravenous (upper panels) and oral administration (lower panels) of equal doses of two totally metabolized drugs. The left panels refer to a drug with a total intrinsic clearance equivalent to an extraction ratio of 0.1 when blood flow equals 1.5 L/min, and the right panels to a drug with an intrinsic clearance equivalent to an extraction ratio of 0.9 (from Ref. [59]) with permission)

ability of the liver to remove drug. Altered hepatic blood flow is therefore not important, making low intrinsic clearance drugs "flow-independent" (Fig. 9, left hand side). Although the efficiency of extraction increases slightly with decreasing flow rates, the resulting change in bioavailability is small. Referring to Fig. 9, bioavailability decreases from 90 percent to 82 percent, therefore, only a nine percent decrease in bioavailability is caused by a 50 percent reduction in hepatic blood flow. Changing bioavailability is therefore not an important factor and even when administered orally, the hepatic clearance, area under the curve and steady-state concentration of a low intrinsic clearance drug remain unchanged when hepatic blood flow is altered.

High intrinsic clearance drugs however, are "flow-dependent", implying that changes in hepatic blood flow should cause parallel changes in hepatic clearance. Figure 9, upper right, illustrates the effect of decreasing hepatic blood flow on the concentration-time curve following an intravenous injection of a high intrinsic clearance drug. In the presence of a sustained decrease in hepatic blood flow, the clearance is decreased and the half-life prolonged. The lower graph depicts the effect of the same decrease in hepatic blood flow on an oral dose of a high intrinsic clearance drug. Again, changes in extraction ratio must be considered to interpret this graph. As Shand has stated [68], reducing hepatic blood flow will increase hepatic extraction and presystemic elimination, so that less drug is available systemically. At the same time, however, the fall in flow reduces hepatic clearance so that the reduced amount of drug that enters the circulation is cleared more slowly, and the area under the curve and steady-state concentration are unchanged. Shand also emphasizes that while the area under the curve is unaltered, the shape of the concentration time curve is changed. The reduced systemic availability leads to lower peak levels, while the fall in clearance prolongs the half-life leading to higher trough concentrations.

Since only a few agents, such as beta blockers, cimetidine and certain vasodilators are known to affect hepatic blood flow, disease states which alter hepatic blood flow, for example CHF, are more frequent causes of alterations in hepatic clearance than drug interactions. Although examples with low intrinsic clearance drugs are limited, there exist several excellent papers detailing the effects of altered hepatic blood flow on high intrinsic clearance drugs [69-71]. Schneck et al. [69] examined the effect of nadolol and propranolol on the clearance of intravenously administered lidocaine (a high intrinsic clearance drug). Both beta-blockers were shown to decrease hepatic blood flow and lidocaine clearance with resulting increases in plasma lidocaine concentrations. Since the intrinsic clearance of lidocaine was not affected by either beta blocker, the authors conclude that both nadolol and propranolol reduce lidocaine clearance by their effects on hepatic blood flow and not by inhibition of lidocaine metabolism.

To demonstrate the effect of changes in hepatic blood flow on the clearance of orally administered high intrinsic clearance drugs, the interaction between propranolol and hydralazine may be examined. Schneck [70] and McLean [71] both reported an increase in peak oral propranolol concentrations when administered with hydralazine. Ideally, the half-life of propranolol should have been shorter due to increased hepatic

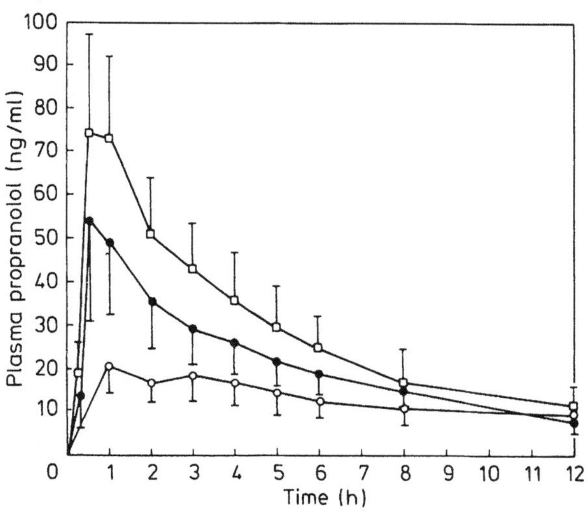

Fig. 10. Effect of hydralazine on plasma propanolol levels. ○——○ 40 mg propranolol; ●——● 40 mg propranolol, 25 mg hydralazine; □——□ 40 mg propranolol, 50 mg hydralazine. Area under the curve increased significantly when both 25 mg and 50 mg of hydralazine were administered concomitantly (from Ref. [70] with permission)

clearance with co-administration of hydralazine. Fig. 10 indicates that this was not the case. Clearance did not appear to be increased by a hydralazine-induced increase in hepatic blood flow in either study. Infact, both studies showed a significant increase in area under the curve (oral) when hydralazine was given in combination with the beta blocker. Schneck attempts to explain this observation by postulating that hydralazine may cause only a transient increase in blood flow, leading to a decreased first pass effect, but dissipating quickly so that overall clearance remains relatively unaffected.

This latter interaction exemplifies the difficulties encountered when examining the mechanisms behind alterations in hepatic drug metabolism. Although examples of pharmacokinetic interactions have been supplied to support these theories, there exist many "metabolic" interactions which do not conform to these principles. Flow dependence versus intrinsic clearance dependence have remained the "standards" for understanding hepatic metabolic interactions, yet continued research is necessary to determine if these mathematical concepts withstand the test of time.

4 Altered Distribution

The volume into which a drug appears to distribute is referred to as the apparent volume of distribution. This is a useful pharmacokinetic parameter that relates the plasma or serum concentration of a drug to the total amount of drug in the body. Therefore, despite its name, the volume of distribution has no direct physiologic meaning and does not refer to a real volume. As a result, if the volume of distribution is reduced, this does not imply that the drug is no longer distributed to a particular part of the body. Infact, following distribution of a drug within the body, any change in the volume of distribution would result in a change in concentration in one or all of the various tissues. The distribution of a drug within the body is dependant on its tissue affinity and solubility in various tissues as well as its ability to traverse the membranes separating one tissue from another.

4.1 Changes in Plasma Protein Binding

Plasma protein binding is one factor which may be altered and can affect the apparent volume of distribution. However, since the plasma protein binding of most drugs is relatively constant with respect to concentration, disease or other drug therapy, its effect on drug elimination or clearance is generally not routinely considered. Nevertheless, changes in plasma protein binding can profoundly affect drug elimination and the interpretation of therapeutic drug concentration determinations.

The protein binding of a drug generally only changes in the following situations:

(i) As a direct result of changes in the concentration of the binding protein. This only occurs in response to disease. For example during a myocardial infarction, alpha-1 acid glycoprotein (AAG) concentrations change dramatically. Since a strong

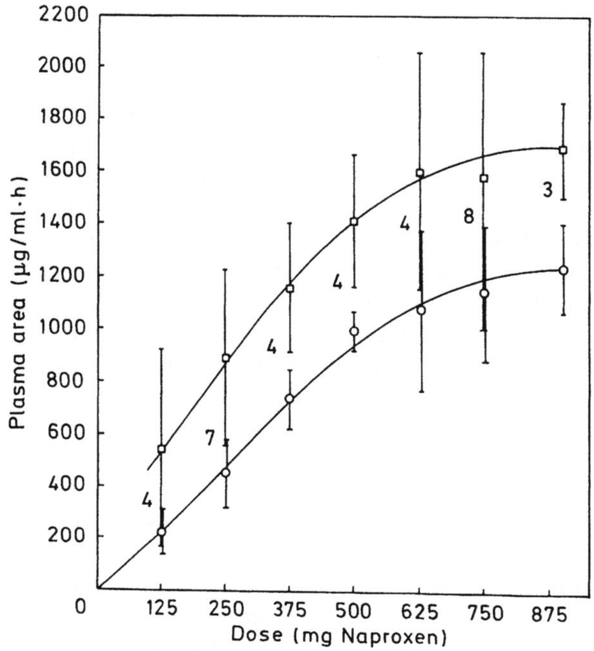

Fig. 11. Effect of naproxen dose on the area under the plasma concentration-time curve for naproxen on day 1 (open circles) and day 5 (open squares) for each naproxen dose. Due to reductions in plasma protein with increasing concentrations (achieved with increasing dose) clearance is increased and area under the plasma concentration-time curve eventually plateaus. Bars indicate 95% confidence intervals, and numerals indicate number of subjects used to calculated means (From Runkel et al. Ref. [75], with permission)

relationship between AAG concentration and lidocaine protein binding has been demonstrated [72,73] the AAG concentration can greatly affect lidocaine clearance [73] and the interpretation of total lidocaine concentration results. Conversely, severely debilitated or malnourished elderly patients, and liver disease or nephrotic syndrome patients can have large reductions in total plasma protein concentrations, primarily albumin. This can cause an increase in the free fraction of some drugs resulting in an increased clearance (eg. phenytoin) [74].

(ii) Where the protein binding of a drug changes as a function of drug concentration. This has been observed to occur with a number of nonsteroidal anti-inflammatory drugs, particularly naproxen [75], and the anticonvulsant valproic acid [76]. When the protein binding is reduced at higher total plasma concentrations, and the drug has a low extraction ratio, then clearance increases with increasing dose. This results in a plateauing of the drug concentration at steady state, even if the dose is further increased [75]. (Fig. 11).

(iii) where a second drug displaces a highly protein bound drug. It should be noted that these "classic" interactions are transient, and unless the displaced drug exerts an immediate, direct effect which is related to the free plasma concentration, a change in effect or response may not even be observed. On theoretical grounds, simple displacement interactions are not likely common, since a number of criteria must be met.

For one drug to displace another and cause a significant interaction, both drugs must compete for common binding sites. A second important criteria is that the displacing drug must be present in a concentration which exceeds the molar concentration of the protein binding sites. Obviously, even if two drugs compete for common binding sites and there is an excess of these binding sites, there would be no change in the protein binding of the first, bound drug. A third important consideration is the degree of protein binding. Displacement of protein bound drug can only be important if there is an important increase in the concentration of free drug. Protein binding displacement interactions are, therefore, only important for drugs which are initially highly protein bound. An arbitary "rule of thumb" is that drugs which are bound less than 75 percent cannot be displaced to an extent which would cause a significant change in the free fraction.

Usually only acidic drugs meet these criteria. Basic drugs, in general, have lower plasma concentrations, and as a result do not fulfill the criteria of adequate molar concentration.

The most common clinical situation in which a drug-drug protein binding displacement interaction is encountered, occurs when a second more avidly bound drug (the displacer) is added to chronic therapy of a less avidly bound drug (the displaced drug). In this situation, the second drug displaces the first and degree of displacement will increase as the concentrations of the displacer increase as steady-state for the displacer is achieved. The immediate effect on the displaced drug is a decrease in the plasma protein binding and an increase in the free concentration (Fig. 12). The free drug is then able to redistribute within the body. Pharmacokinetically, this will result in an increase in the volume of distribution. Evaluation of the initial phase of such an interaction can become complicated if you wish to consider the absorption and distribution rates of the displacing drug and the distribution rate of the displaced drug. If absorption occurs faster than distribution, the high pre-distribution concentration of displacing drug will further increase the plasma protein binding displacement, enhancing the interaction. Clinically, there may be a change in the drug response. However, this will depend on the amount of drug which actually redistributes and the volume into which it can disperse. If redistribution is significant and the concentration at the site of action is reduced, then the observed effect might be a reduction in clinical response.

The fate of the increased, unbound drug is dependant on the efficiency with which the liver can remove drug from plasma. For a highly extracted drug, hepatic clearance is unaffected by the fraction which is unbound. This is termed "non-restrictive" clearance [59]. Therefore, drug displacement should not affect total drug clearance or average steady-state drug concentration, because, by definition, the extraction ratio is greater than the free-fraction of drug (i.e. the liver very efficiently eliminates drug). However, because the displacement increases the free drug concentration, this drug can now re-distribute, increasing the volume of distribution. As a result, half-life will be prolonged. The end result of such an interaction should be a permanently altered free drug concentration since no compensatory increase in elimination rate can occur. However, displacement interactions involving drugs which are non-restrictively eliminated may not be clinically relevant, since these drugs are highly extracted by the liver and have widely variable plasma concentrations following oral administration due to first pass effect. Therefore, only large effects would show against this background of large individual variation [78]. Furthermore, displacement interactions have not yet been described for "non-restrictively" eliminated drugs.

For a drug which is poorly extracted, drug clearance is dependent on the free concentration of the drug [59]. Therefore, following displacement, the increase in free fraction causes a proportional increase in the drug clearance and the total drug concentration declines to a new steady state. The increased elimination rate will be observed as a shorter half-life and so the time course to achieve the new steady-state will be much shorter than would generally be predicted. During the period between initial displacement and achievement of the new steady-state, the increased free drug will distribute throughout the body. It is at this time that any change in drug response will be observed. Once this new steady state is achieved, the concentration of free

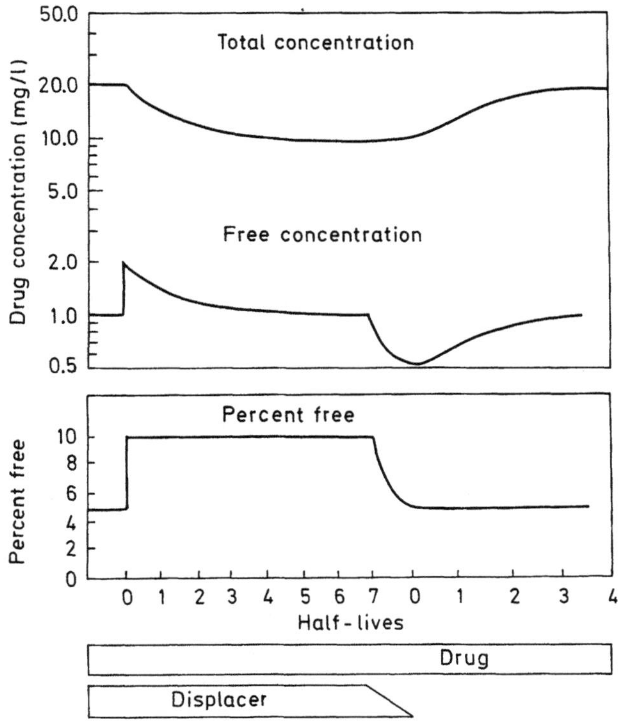

Fig. 12. Effect of administration of a displacing drug on the unbound and total plasma concentrations of a constantly infused low extraction drug. Prior to bolus administration of the displacing drug, the drug of interest is about 95% bound to plasma proteins. Immediately following administration of the displacer, plasma protein binding is reduced to 90%, and the free concentration increases from 1 mg/L to 2 mg/L. The increased free drug is then eliminated and eventually a new steady state is reached. The free concentration has returned to 1 mg/L, identical to the pre-displacement concentration. When the displacing drug is removed, all parameters eventually return to pre-displacement conditions (Adapted from Rowland et al. Ref. [77])

drug will be identical to its pre-displacement concentration. However, because of the greater free fraction, there is a lower total concentration. There is a direct relationship between the increase in the free fraction and the increase in total plasma clearance (Fig. 12).

At this new steady-state, clinical response will be identical to the pre-displacement response. Now, one of the most serious potential consequences of any protein binding displacement interaction can occur, as a result of misinterpretation of a drug concentration report. Although the initial displacement may have resulted in an increased, but transient enhanced response, it quickly corrected itself. If, however, the physician now requests a blood be drawn to ensure the concentration is in the usual therapeutic range, the report may indicate a "low" total plasma concentration. Interpretation of this concentration should be done with great caution, since, although the total concentration is low, the free concentration is identical to the pre-displacement concentration and may well be within an acceptable range. Increasing the dose to achieve the more acceptable total therapeutic drug concentration would increase the risk of toxic side-effects.

Often, interactions involving protein binding displacement involve other mechanisms as well. For example, although phenylbutazone displaces warfarin from plasma albumin in-vitro and in vivo, it is also known to inhibit the biotransformation of the S-isomer, while enhancing the elimination of the less potent R-isomer [79]. The chronic hypoprothrombinemic effect that this interaction produces is due to the inhibition of the S-isomer elimination, not by protein binding displacement. [79]. It also appears that valproic acid can displace plasma protein bound phenytoin [80–82] as well as inhibit the metabolism of phenytoin [82]. The addition of valproic acid to a stable phenytoin regimen increases the phenytoin free fraction from about 9–15 percent [80–82]. This displacement is associated with both an increased systemic clearance and apparent volume of distribution for total drug [82]. However, it has also been demonstrated that the free drug clearance of phenytoin is reduced [82] suggesting inhibition of hepatic metabolism. Chloral hydrate-warfarin is an example of a simple protein binding displacement interaction [83]. Chloral hydrate is metabolized in part to trichloralacetic acid, which can displace warfarin from plasma proteins. The increased free fraction of warfarin can distribute throughout the body, but at the same time warfarin metabolism is increased and a new steady-state is being achieved. Whether a significant increase in prothrombin time is actually observed is questionable. It is known that the inhibition of vitamin K dependant factor synthesis is transiently intensified during the period immediately following displacement. However, clotting factors already circulating in the blood must be eliminated before a clinical change in prothrombin time can be observed. Since these factors have relatively long half-lives [84], a change in the prothrombin time may not be observed. While it is possible that, for some very sensitive individuals, a change in PT will be observed, it has been shown that warfarin requirements are identical to predisplacement requirements within four days of the initial protein-binding displacement [85].

4.2 Changes in Tissue Binding

Drugs can bind to tissue proteins as well as plasma proteins. All of the criteria required for a significant plasma protein binding displacement interaction must also be present if a tissue protein binding displacement interaction is to occur. Furthermore, redistribution will also occur, but now it occurs in the reverse direction away from tissue sites. If plasma protein binding does not also change, any drug redistributing back into plasma will increase the plasma concentration. If plasma is the reference tissue, then after analysis of the plasma concentration this situation appears to be no different than having a proportionately increased dose. Any change in clinical response will be dependant on the changes in the free concentration at the site of action.

The digoxin-quinidine interaction is an example of such a tissue binding displacement interaction [86-90]. Fifty percent of the total body store of digoxin is located in skeletal muscle [88]. Quinidine is reported to displace tissue bound digoxin from skeletal muscle [88]. The displacement of tissue bound digoxin reduces the volume of distribution and produces approximately a two fold increase in the serum digoxin concentration, although the increase is quinidine dose dependant. There is no change in digoxin plasma protein binding. It has been reported that the combination of digoxin and quinidine reduces the positive inotropic effect of digoxin [89]; however, others have demonstrated that quinidine intensifies typical digoxin electrophysiologic effects [90]. There also appears to be an increase in non-cardiac adverse reactions (eg. gastrointestinal side effects).

5 Altered Renal Excretion

The kidney is a major organ involved in the elimination of drugs from the body. Clearance of a drug or metabolite by the kidney can occur by a number of mechanisms. In addition to direct renal metabolism, the primary methods of elimination are glomerular filtration, tubular secretion, or both. Tubular reabsorption, which is defined as reabsorption from the lumen of the tubule into the tubular cell in a direction opposite to secretion, is not truly a method of elimination, but, is also a potential site for some drug-drug interactions.

5.1 Renal Reabsorption

Inhibition or interference with renal tubular reabsorption occurs with a number of organic acids. Probenecid, phenylbutazone and high dose salicylate can produce uricosuria by inhibiting uric acid reabsorption into the proximal tubule. Aminoglycosides are also reabsorbed into proximal tubular cells following an initial binding to the brush border [91,92]. Animal studies have demonstrated that a number of polyamines including N-methyl nicotinamide [93] and quinine [94] can reduce cortical accumulation [94] and result in greater urinary recovery of the aminoglycoside [93]. Limited data produced in our laboratory in five subjects has also demonstrated that a single 1200 mg dose of quinine can increase 24 hour urinary recovery of gentamicin following a single 80 mg dose of the antibiotic.

5.2 Glomerular Filtration

Direct interference with glomerular filtration of one drug by another is unknown. Protein binding displacement can result in an increased free concentration. This increased free drug can then be directly filtered, resulting, theoretically, in increased urinary excretion and recovery rates. However, this has not been well documented.

Reductions in glomerular filtration rate (GFR) through inhibition of prostaglandin biosynthesis is the proposed mechanism whereby many nonsteroidal anti-inflammatory drugs may cause a drug-drug interaction. For example, it has been shown that indomethacin increases the standard 12 hour steady-state plasma concentration of lithium and decreases lithium renal clearance in both normal volunteers and psychiatric patients. The increase in the plasma concentration averaged 59 percent in patients and 30 percent in normals [95].

5.3 Secretion

The pars recta (straight segment) of the proximal tubule actively secretes a variety of organic acids and bases into the tubular lumen. Although the pathways for acids and bases appear to be separate, within each pathway competition for secretion can occur. The best known example of a drug-drug interaction involving inhibition of secretion is one which is often used clinically to reduce the elimination of penicillin. Probenecid which is actively secreted, can compete with penicillin and some cephalosporins to reduce the renal clearance of these antibiotics. This, of course, reduces the elimination rate of the antibiotic, prolonging its action and often reducing the need for a more frequent dosing regimen.

It is also well known that loop diuretics, like furosemide, must reach the tubular lumen to be effective [96]. Furosemide is actively secreted via the organic acid transport pathway. Pretreatment with probenecid can reduce the rate at which furosemide is initially delivered to the tubular lumen [96,97]. As a result, there is initially a decreased response to furosemide administration in the first 60–90 minutes. However, because of early initial reductions in renal clearance which result in higher plasma furosemide concentrations, there is an increased diuretic response to later times. This later response is sufficient to eliminate any statistical differences in mean eight hour values for sodium and volume excretion [97].

Indomethacin, another organic acid, can also attenuate the natriuretic, hemodynamic and renin stimulating effects of furosemide. Although the mechanism of this interaction is often reported as inhibition of prostaglandin synthesis, an alternative explanation is that indomethacin competes with the active secretion of furosemide into the tubular lumen. Flolich et al. [98] has demonstrated an 18 percent reduction in furosemide elimination when co-administered with indomethacin. Smith et al. [99] have also studied this interaction and similarly found a reduced renal clearance clearance of furosemide. However, although both investigators [98,99] demonstrated that indomethacin affects the disposition of furosemide, the changes were minimal and cannot entirely explain the marked effect that indomethacin has on furosemide response.

5.4 *Renal Metabolism*

Renal elimination of drugs is generally regarded to occur solely through direct urinary excretion. However, the kidney can also metabolize many drugs. Acetaminophen, in overdose, is well known to cause liver damage, but can also cause renal failure. The renal failure is a result of proximal tubular necrosis which occurs as a result of acetaminophen metabolism involving similar P450 pathways to those found in the liver [100].

Therefore, metabolic inhibition and induction, reviewed earlier, can also apply to renal metabolism of drugs. However, rarely is the kidney responsible for a large percentage of the overall metabolic biotransformation occurring for any one drug.

Metabolism, specific to the kidney, however, does occur. Enzymes capable of metabolizing drugs can be found in the brush border of the renal proximal tubule. Imipenem is a beta-lactam carbapenem antibiotic, which can be inactivated by the opening of the beta-lactam ring by dehydropeptidase-1 [101]. As a result, both filtered and secreted imipenem can be metabolized. To reduce this metabolism, cilastatin, a dipeptidase inhibitor, can be given with imipenem to increase the amount of active imipenem appearing in the urine. When imipenem and cilastatin are co-administered, imipenem is effective in the treatment of urinary tract infection caused by susceptible organisms, since up to seventy percent of the administered dose of the active parent compound appears in the urine (Fig. 13). This is considerably greater than the five to 40 percent of the dose appearing unchanged in the urine when imipenem is administered alone [102].

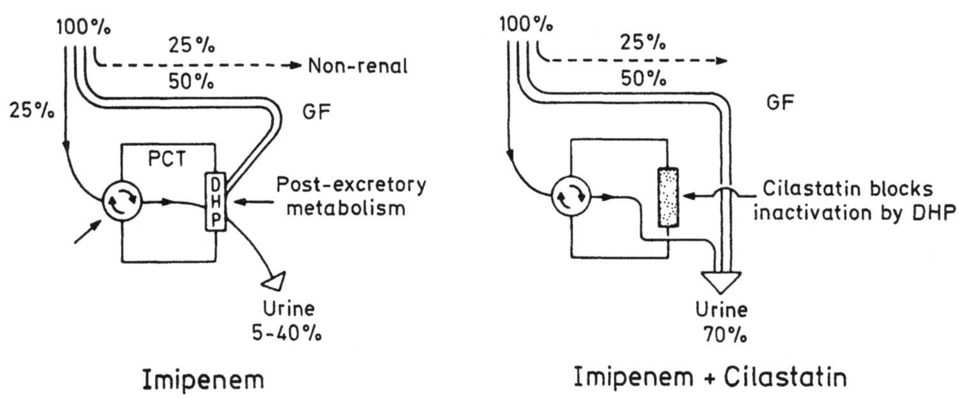

Fig. 13. About 25% of Imipenem is eliminated by nonrenal mechanisms, 50% by glomerular filtration (GF), and 25% by active tubular secretion. However, both filtered and secreted drug is metabolized by the dehydropeptidase (DPH) enzyme found on the brush border of the proximal convoluted tubular (PCT) cell, so that only 5–40% of unchanged imipenem reaches the urine. Co-administration of cilastin and imipenem (right panel) inhibits the dehydropeptidase enzyme, allowing most of the filtered and secreted imipenem to escape into the urine unchanged (from Barza, Ref. [102], with permission)

6 Clinical Significance of Drug Interactions

The clinical significance of pharmocokinetic drug interactions frequently depends upon the summation of many variables that govern a patient's response to a drug. Vessel [103] has documented at least 20 factors that alone affect hepatic drug metabolism. A majority of pharmacokinetic drug interaction studies are carried out using healthy volunteers under ideal conditions that attempt to control for many of these variables; thus co-existing pathologic, nutritional, genetic, or environmental factors could magnify or lessen a pharmacokinetic interaction demonstrated under experimental conditions. In general, as stated by Shand et al. [104], the significance of pharmacokinetic drug interactions that decrease by any means the amount of drug available to the receptor site is proportional to the severity of the underlying disease. Conversely, the importance of interactions that increase the amount of drug in the body is proportional to the therapeutic index and inherent toxicity of the drug. Some of the other factors that govern the clinical significance of a drug interaction include the following situations.

6.1 *Likelihood of Co-administration*

To result in a clinically significant interaction, the drugs have to have a reasonable chance of being prescribed together in practice. For example, propranolol was recently shown to inhibit the metabolism of theophylline [105]. However, the likelihood of propranolol, which can exacerbate an asthmatic attack, being prescribed to a patient taking theophylline is very small. On the other hand, the impact of interactions with cimetidine [106], one of the most frequently prescribed drugs world-wide is much greater. The clinical significance of any drug interactions with non-prescription drugs is probably enhanced by the limited professional input into or control over their use in the unsuspecting patient.

6.2 *Inherent Properties of the Drug or Disease State*

Interactions involving drugs that have a narrow therapeutic index, result in a severe adverse effect in overdosage, or are used to treat life-threatening diseases or unique medical conditions are always potentially significant.

A high percentage of drug interaction reports accumulated in texts on the subject are accounted for by drugs such as warfarin, digoxin, anti-convulsants, and anti-dysrrythmics [23]. All of these compounds have a narrow therapeutic index, in which the range of serum concentrations or required dosage over which a drug is effective without producing toxicity, is small.

Antineoplastic agents by nature of their pharmacologic action are generally toxic even in their usual clinical setting. Any drug that prolongs their residence time in the body is likely to produce severe toxicity. Thyss et al. [107] recently reported that simultaneous administration of ketoprofen, a non-steroidal anti-inflammatory agent, with methotrexate to four patients probably lead to the death of three of them;

the mechanism proposed was decreased renal excretion of the methotrexate with resultant toxic serum methotrexate levels.

The reduced effectiveness of oral contraceptive agents by concomitant use of rifampin or phenobarbital [108] has lead to a clinical outcome that is not altered by simple discontinuation of the enzyme inducing agent.

6.3. *Pharmacokinetic Factors*

Several pharmacokinetic factors may determine the clinical significance of drug interactions. The large intersubject variability in drug disposition may be one of the more important factors. We have shown that the decrease in phenytoin clearance by cimetidine varied from 0–25 percent in eight subjects [12]. This suggests that some epileptic patients could receive this combinations without developing signs of toxicity. On an average, those patients with high initial drug clearances relative to the general population are most likely to exhibit significant percent changes in changes in clearance when an enzyme inhibitor is added. Conversely, patients with low clearances may have the most dramatic response to enzyme induction.

The dose of the interacting drug may also determine the clinical significance of the interaction in two ways. Firstly, the magnitude of the result of enzyme induction [109], enzyme inhibition [12] and intestinal adsorption [39] has been shown to increase with increasing doses of the interacting drug; secondly, competitive hepatic enzyme inhibition with drug toxicity may occur early on during the administration of an enzyme-inducing drug whose eventual effect will be decreased drug efficacy [110].

A significant pharmacokinetic interaction may not always lead to a clinical one of a similar magnitude. For example, Bodeman et al. [111] showed that cimetidine plasma concentrations were significantly lowered by simultaneous administration of antacids in the fasting state. However, the lower plasma cimetidine levels achieved were still greater than that required to suppress acid secretion by 50 percent and would be unlikely to clinically affect ulcer healing.

The clinical significance of a pharmacokinetic drug interaction resulting from a change in motility of the upper gastrointestinal tract will depend upon several pharmacologic and pharmacokinetic properties of the affected drug. For example, a drug with anticholinergic effects that delay gastric emptying would more likely affect a drug, such as an oral analgesic, whose therapeutic effectiveness is dependent upon a relatively rapid absorption rate; this type of interaction would have less of a clinical significance for a drug such as phenytoin, whose pharmacologic effect is dependent upon maintaining a certain steady-state serum concentration rather than a specific peak serum concentration.

6.4 *Pharmacogenetic Factors*

Genetic influences could have a role to play in drug interactions involving oxidative drug metabolism, a major determinant of the efficacy, duration of action and toxicity of many drugs. Six to 10 percent of caucasians have an impaired ability to eliminate drugs such as debrisoquine, metoprolol and propranolol [112]. Presently, at least, there is no simple means of determining those patients who are "poor metabolizers" and yet, theoretically, these patients may be at greater risk for significant pharmacokinetic drug interactions involving hepatic enzyme induction or inhibition.

6.5 *Other Factors*

For some interactions, corrective action taken as a result of the co-administration of the two drugs can be more clinically significant than the interaction itself. This is especially true of protein binding displacement interactions and may occur when the dose of a drug, such as phenytoin or valproic acid, whose therapeutic effect is monitored by the measurement of serum concentrations, is increased because an interacting drug has lowered the reported total serum concentration to sub-therapeutic concentrations; infact, the free serum concentration, which determines pharmacologic effect, has not changed and the unnecessary dosage increase could lead to toxicity.

7 Outlook

Like all adverse drug effects, every attempt should be made to predict and avoid drug-drug interactions. The prevention of pharmacokinetic drug interactions should be the responsibility of clinical practitioners, clinical investigators, the pharmaceutical industry and drug regulatory agencies. Physicians and pharmacists must play the role of the "prepared observer" [113]. The co-administration of drugs with those possessing a narrow therapeutic index must always be untertaken with caution or suspicion, even if no published documentation is available. A drug that produces a pharmacokinetic drug interaction by a specific mechanism should be assumed to produce that interaction with other drugs that are similarly absorbed, metabolized or excreted as the index drug until proven otherwise.

Clinical investigators must be willing to follow up case reports [114] of drug interactions with pharmacokinetic studies to determine its mechanism and potential clinical significance [19].

Drug regulatory agencies must demand and the pharmaceutical industry provide pre-marketing data on new drugs that may produce a pharmacokinetic drug interaction based on previously published data on similar drugs. Examples where this has recently been carried out exist: the effect of H_2-antagonists, in the investigational stage, on hepatic drug metabolism [115] and the effect of non-steroidal anti-inflammatory agents on the pharmacokinetics and pharmacodynamics of new antihypertensives [116].

8 Acknowledgement

The authors wish to express gratitude to their colleague, Dr. Brian Hardy, for a constructive critique of this manuscript and to Sonia Amair for her dedicated clerical assistance.

9 References

1. Robinson, D. S.: J. Urol. *113*, 100 (1975)
2. Hurwitz, N., Wade, O. L.: Br. Med. J. *1*, 531 (1969)
3. Crooks, J. et al.: The clinical significance and importance of drug interactions, in: Drug Interactions (ed.) Grahame-Smith, D. G., p. 3, Baltimore, University Park Press (1977)
4. Talley, R. B., Laventurier, M. F.: J. Amer. Med. Assoc. *220*, 1287 (1972)
5. Steward, R. B., Cluff, L. E.: Jons Hopkins Med. J. *129*, 319 (1971)
6. McLauglin, J. E. et al.: Lancet *1*, 261 (1971)
7. Watkins, J. et al.: Br. Med. J. *281*, 702 (1980)
8. Fichtl, B. and Doering, W.: Clin. Pharmacokinet. *8*, 137 (1983)
9. Brown, D. D. and Juhl, R. P.: New Engl. J. Med. *295*, 1034 (1976)
10. Levy, G. et al.: New Engl. J. Med. *293*, 323 (1975)
11. Van Der Meer, J. W. M., et al.: J. Antimicrob. Chemother. *6*, 552 (1980)
12. Bartle, W. R. et al.: Clin. Pharmacol. Ther. *33*, 649 (1983)
13. Somogyi, A. et al.: Eur. J. Clin. Pharmacol. *25*, 339 (1983)
14. Oppenheimer, J. H. and Tavernetti, R. R.: J. Clin. Invest. *41*, 2213 (1962)
15. Lander, C. M. et al.: Clin. Exp. Neurol. *14*, 184 (1977)
16. O'Reilly, R. A. and Levy, G.: J. Pharm. Sci. *59*, 1258 (1970)
17. O'Reilly, R. A. et al.: J. Clin. Invest. **65**, 746 (1980)
18. Lindenbaum, J. et al.: New Engl. J. Med. *305*, 789 (1981)
19. Schwartz, J. et al.: Drug Intell. Clin. Pharm. *17* (1983)
20. Neuvonen, P. J. et al.: Brit. J. Clin. Pharmacol. *2*, 94 (1975)
21. Kirch, W. et al.: Clin. Pharmacol. Ther. *39*, 202 (1986)
22. Shinn, A. F., Shrewsbury, R. P.: Evaluations of Drug interactions, St. Louis—Toronto—Princeton, C. V. Mosby, 1985[3]
23. Hansten, P. D.: Drug Interactions. Philadelphia, Lea & Febiger, 1985
24. Melander, A.: Clin. Pharmacokinet. **3**, 337 (1978)
25. Bayne, W. et al.: Clin. Pharmacol. Ther. *32*, 270 (1982)
26. Wilson, J. P.: Gut *8*, 618 (1967)
27. Schoder, H. and Campbell, D. E. S.: Clin. Pharmacol. Tner. *13*, 539 (1972)
28. Heading, R. C. et al.: Brit. J. Pharmacol. *47*, 415 (1973)
29. Welling, P. G.: Clin. Pharmacokinet. **9**, 404 (1984)
30. Beerman, B. and Groschinsky-Grind, M.: Eur. J. Clin. Pharmacol. *12*, 297 (1977)
31. Manninen, V. et al.: Lancet *1*, 398 (1973)
32. Hurwitz, A. et al.: Gastroenterol. *71*, 268 (1976)
33. Bartle, W. R. et al.: Amer. J. Hosp. Pharm.
34. van Hees, P. A. M. et al.: Gut *20*, 300 (1979)
35. Lindenbaum, J. et al.: Gastroenterol. *71*, 390 (1976)
36. Barr, W. H. et al.: Clin. Pharmacol. Ther. *12*, 779 (1971)
37. Van Der Meer, J. W. M. et al.: J. Antimicrob. Chemother. 1980; 6: 552–4.
38. Feldman, S. and Carlstedt, B. C.: J. Amer. Med. Assoc. *227*, 660 (1974)
39. Neuvonen, P. J.: Clin. Pharmacokinet. *7*, 465 (1982)
40. Neuvonen, P. J. and Elonen, E.: Eur. J. Clin. Pharmacol. *17*, 51 (1980)
41. Parsons, R. L. and Paddock, G. M.: J. Antimicrob. Chem. *1* (Suppl.) 59 (1975)
42. Kauffman, R. E. and Azarnoff, D. L.: Clin. Pharmacol. Ther. *14*, 886 (1973)
43. Caldwell, J. H. et al.: J. Clin. Invest. *50*, 2638 (1971)
44. Neuvonen, P. J.: Drugs *11*, 45 (1976)
45. Lindenbaum, J. et al.: New Engl. J. Med. *305*, 789 (1981)
46. Curry, S. H., Aburawi, S. M.: Biopharmaceut. Drug Metab. *6*, 235 (1985)
47. Shaw, J. O., Moser, K. M.: Chest *68*, 75 (1975)
48. George, C. F.: Clinical Pharmacok. *4*, 433 (1979)
49. Branch, R. A. et al.: Drug Metabol. & Disposition *1*, 687 (1973)
50. Powell, J. R., Donn, K. H.: Amer. J. Med. *77* (Suppl 5B), 57, (1984)
51. Christensen, L. K., Skovsted, L.: Lancet *2*, 1397 (1969)
52. O'Reilly, R. A.: Ann. Intern. Med. *78*, 73 (1973)

53. Brown, M. W. et al.: Hepatology *4*, 1036 (1984)
54. Kramer, P., McClain, C.: New Engl. J. Med. *305*, 1262 (1981)
55. Vessell, E. S., Page, J. P.: J. Clin. Invest. *48*, 2202 (1969)
56. Zilly, W. et al.: Eur. J. Clin. Pharmacol. *9*, 219 (1975)
57. Hanssen, J. M. et al.: Clin. Pharmacol. Ther. *12*, 539 (1971)
58. Huffman, D. H. et al.: Pharmacology *10*, 338 (1973)
59. Wilkinson, J. R., Shand, D. J.: Clin. Pharmacol. & Ther. *18*, 377 (1975)
60. MacDonald, M. G. et al.: Clin. Pharmacol. Ther. *10*, 480 (1969)
61. O'Reilly, R. A.: Ann. Intern. Med. *81*, 337 (1974)
62. Renton, K. W. et al.: Clin. Pharmacol. Ther. *30*, 422 (1981)
63. Laforce, C. F. et al.: J. Pediatr. *99*, 153 (1981)
64. Grygiel, J. J. et al.: Eur. J. Clin. Pharmacol. *26*, 335 (1984)
65. DeAngelis, C. et al.: Clin. Pharm. *2*, 563 (1983)
66. Herman, J. et al.: Brit. J. Clin. Pharmacol. *16*, 565 (1983)
67. Vestal, R. E. et al.: Clin. Pharmacol. Ther. *25*, 19 (1979)
68. Shand, D. G.: Clinical Pharmacokinetics, in: Recent Advances in Clinical Pharmacology (eds.) Turner, P., and Shand, D. G. P. 6, Churchill Livingstone (1978)
69. Schneck, D. W., Davis, D., Vary J. et al.: Clin. Pharmacol. Ther. *36*, 584 (1984)
70. Schneck, D. W.,.Vary, J. E.: Clin. Pharmacol. Ther. *35*, 447 (1984)
71. McLean, A. J. et al.: Clin. Pharmacol. Ther. *27*, 726 (1980)
72. Routledge, P. A.: Clin. Pharmacol. Therap. *27*, 347 (1980)
73. Routledge, P. A.: Clin. Pharmacol. Therap. *30*, 154 (1981)
74. Hayes, M. J.: Br. J. Clin. Pharmacol. *2*: 73 (1975)
75. Runkel, R.: Clin. Pharmacol. Ther. *15*, 261 (1974)
76. Brewster, D.: Clin. Pharmacol. Therap. *27*, 76 (1980)
77. Rowland, M., and Tozer, T. N. in: Clinical Pharmacokinetics: Concepts and Applications. Pg. 255. Philadelphia, Lea and Febiger (1980)
78. Shand, D. G. et al.: Pharmacokinetic Drug Interactions, in: Handbook of Experimental Pharmacology. (eds.) Eichler., C., Farah, A., Herken, H. and Welch, A. D. pg. 272. Berlin Springer-Verlag (1975)
79. Lewis, R. J.: J. Clin. Invest. *53*, 1607 (1974)
80. Friel, P. N.: Therap. Drug Monitoring *1*: 243 (1979)
81. Dahlqvist, R.: Br. J. Clin. Pharmacol. *8*: 547 (1979)
82. Perucca, E.: Clin. Pharmacol. Therap. *28*: 779 (1980)
83. Sellers, E. M.: New Engl. J. Med. 283, 827 (1970)
84. Biggs, R., Denson, K. W. E.: Br. J. Haematol. *9*, 532 (1963)
85. Boston Collaborative Drug Surveillance Program. New Engl. J. Med. *286*, 53 (1972)
86. Doering, W.: New Engl. J. Med. *301*, 400 (1979)
87. Leahey, E. B.: Am. J. Cardiol. *48*, 1141 (1981)
88. Schenck-Gustafsson, K.: New Engl. J. Med. *305*, 209 (1981)
89. Steiness, E.: Clin. Pharmacol. Ther. *27*, 791 (1980)
90. Leahey, E. B.: Arch. Intern. Med. *139*: 519 (1979)
91. Just, M.: Nauyn-Schmiedeberg's Arch. Pharmacol. *300*: 57 (1977)
92. Just, M.: Nauyn-Schmiedeberg's Arch. Pharmacol. *300*: 67 (1977)
93. Jerauld, R.: Antimicrob. Agents Chemoth. *13*: 893 (1978)
94. Whelton, A.: Kid. Intern. *6*, 131 (1974)
95. Frolich, J. C.: Br. Med. J. *1*: 1115 (1979)
96. Odlind, B.: Clin. Pharmacol. Therap. *27*: 784 (1980)
97. Benet, L.: J. Pharmacokin. Biopharm. *9* (3), 1 (1979)
98. Frolich, J. C.: Circ. Res. *39*: 447 (1976)
99. Smith, D. E. et al.: J. Pharmacokin. Biopharm. *7* (3), 265 (1979)
100. Mohandas, J.: Res. Commun. Chem. Path. Pharmacol. *34*: 69 (1981)
101. Kahan, F. M.: J. Antimirob. Chemother. *12*: 1 (1983)
102. Barza, M.: Ann. Intern. Med. *103*, 552 (1985)
103. Vessell, E. S.: Clin. Pharmacol. Ther. *31*, 1 (1982)

104. Shand, D. G. et al.: Pharmacokinetic Drug Interactions, in: Handbook of Experimental Pharmacology (eds.) Eichler, C., Farah, A., Herken, H., Welch, A. D., pg. 95, Berlin Springer Verlag (1975)
105. Conrad, K. A., Nyman, D. W.: Clin. Pharmacol. Ther. *28*, 463 (1980)
106. Griffin, J. W. et al.: Amer. J. Med. *77*, (Suppl 5B) 85, (1984)
107. Thyss, A. et al.: Lancet *1*, 256 (1986)
108. D'Arcy, P. F.: Drug Intel. Clin. Pharm. *20*, 353 (1986)
109. Ohnhaus, E. E. et al.: Eur. J. Clin. Pharmacol. *24*, 247 (1983)
110. Levine, M., Sheppard, I.: Clin. Pharm. *3*, 200 (1984)
111. Bodeman, G. et al.: Lancet *1*, 444 (1979)
112. Lennard, M. S. et al.: Clin. Pharmacokinet. *11*, 1 (1986)
113. Melmon, K. L., Nierenberg, D. W.: New Engl. J. Med. *304*, 723 (1981)
114. Bartle, W. R.: Arch. Intern. Med. *140*, 985 (1980)
115. Sambol, N. C. et al.: Clin. Pharmacol. Ther. *39*, 225 (1986)
116. Bartle, W. R. et al.: Acta Pharmacol. Toxicol. Suppl. V, 181 (1986)

Author Index Volumes 1–5

The volume numbers are printed in italics